# Can We Count All the Faces?
## Exploring Simplicial Complexes

Goethe

# Contents

# Introduction

Simplicial complexes are an important tool in the study of topological spaces and manifolds as triangulations of these spaces. A classical problem in algebraic combinatorics and discrete geometry is the face enumeration of simplicial complexes, meaning the classification of the $f$-vectors of simplicial complexes and also of special classes among them. Studying the face numbers of simplicial complexes depending on their topological or combinatorial properties led to a great variety of research, see for example [KN16] for an overview.

A complete characterization giving bounds for the face numbers of simplicial complexes, which are now also known as the Kruskal-Katona inequalitites, was given by Kruskal [Kru63] and Katona [Kat68]. In addition to this general result, one can try to characterize the $f$-vectors for more specific classes of simplicial complexes, such as boundaries of simplicial polytopes, the more general class of simplicial spheres, or balanced complexes. The famous $g$-conjecture, originally raised by McMullen for the boundary complexes of simplicial polytopes in 1971 [McM71], asks for a characterization of the face numbers for simplicial spheres. For boundary complexes of simplicial polytopes the conjectured characterization was proven by 1980. Billera and Lee [BL81] showed the sufficiency of the claimed conditions on the $f$-vector, while Stanley [Sta80] proved the necessity, resulting in the $g$-theorem. Recently, Adiprasito [Adi18] announced a proof of the conjecture in full generality for simplicial spheres.

Stanley showed in his part of the proof of the $g$-theorem by using the Hard Lefschetz Theorem for toric varieties, that the $g$-vector of a boundary complex of a simplicial polytope is an $M$-sequence. From an algebraic point of view, the result of Stanley states that the boundary complex of a simplicial polytope has the *strong Lefschetz property* over a field of characteristic zero.

Motivated by its importance for the $g$-theorem, Lefschetz properties were shown for several other classes of simplicial complexes, for example strongly edge decomposable simplicial complexes [Mur10]. Also the behavior of the Lefschetz property under certain operations on the simplicial complex has been studied, e.g., for join, connected sum and stellar subdivision [BN10]. Another class for which a certain Lefschetz property, called almost strong Lefschetz property, holds, are barycentric subdivisions of shellable simplicial complexes [KN09].

The barycentric subdivision is maybe the most prominent example of a *subdivision* of simplicial complexes. In 1992, Stanley [Sta92] and Kalai asked about the behaviour of the $h$-vector of simplicial complexes under subdivisions, in particular they asked if the entries of the $h$-vector of Cohen–Macaulay complexes increase under certain types of subdivisions. To answer this question, Stanley introduced the so-called *local h-vector*, and could answer the question in the affirmative for so-called *quasi-geometric subdivisions*. To do so, Stanley showed that the local $h$-vector is symmetric in general, and furthermore nonnegative for quasi-geometric subdivisions. Another question arising is the characterization of local $h$-vectors of subdivisions, which in the aftermath was answered for several classes of subdivisions in [Cha94] and [JKMS19]. Since the local $h$-vector is symmetric, one can consider and study its $\gamma$-vector, the so-called local $\gamma$-vector. The nonnegativity of this vector can be connected to Gal's conjecture [Gal05], stating that the $\gamma$-vector of flag spheres is nonnegative. Considering the nonnegativity of the local $h$-vector, one possible question is if its entries can be associated to the number of certain (combinatorial) objects. For example Stanley [Sta92] showed, that the local $h$-vector of the barycentric subdivision is closely connected to the number of derangements, and also for other subdivisions as the $r$-th edgewise subdivision [Ath16a] and the interval subdivision [AS13] combinatorial interpretations have been proven, connecting this invariant to certain statistics on permutations. Also for local $\gamma$-vectors of the before mentioned subdivisions one can give combinatorial interpretations, see [AS12], [Ath16a], [AN18]. One can also ask for other nice properties for these invariants, such as unimodality and log-concavity, which are connected to the real-rootedness of the associated polynomials. For example the real-rootedness of the the local $h$-polynomial of the barycentric subdivision was shown by Zhang [Zha95].

Brenti and Welker [BW08] gave a combinatorial characterization for the $h$-vector transformation of the barycentric subdivision of a complex with nonnegative $h$-vector, and also proved the real-rootedness of its $h$-polynomial. Analogous results have been shown for the edgewise subdivision, see [Joc18], [Ath20b], and the interval subdivision [AN20].

The aim of this book is to study these question for a certain subdivision of simplicial complexes, called *antiprism triangulation*. This triangulation is closely related to the barycentric subdivision in the following sense: the barycentric subdivision of a simplex $\Delta$ can be constructed by inserting a vertex in the interior of $\Delta$ and con-ing over its proper faces, which have been barycentrically subdivided by induction, while the antiprism triangulation can be constructed by inserting another simplex of the same dimension in the interior of $\Delta$ and joining each nonempty face of that simplex with the antiprism triangulation of the complementary face of $\Delta$. The book is based on the results in [ABJK20], the main results can be found in the third and fourth chapter.

## Summary of the book

In the first chapter we start with some basic definitions on simplicial complexes, permutation statistics and polynomials. First, we define simplicial complexes, some subcomplexes and constructions, and important invariants. We furthermore introduce the Stanley-Reisner ring and several special classes of simplicial complexes, and state the Kruskal-Katona inequalities. In the second part, we state several statistics on permutations, and define the Eulerian numbers as well as statistics on $r$-colored permutations. In the last part of this chapter we review different properties of polynomials and their connections.

The second chapter gives an overview on subdivisions of simplicial complexes. In the first part, subdivisions of simplicial complexes as well as the local $h$-vector, one of the main tools in this area, are defined and some general results on it are stated. Afterwards we shed some light on the connection of some well known conjectures such as Gal's conjecture to conjectures on invariants of subdivisions. We conclude this chapter by giving several explicit examples of subdivisions, and collecting known results on the combinatorics of these examples.

The third chapter contains the first main result of this book, the real-rootedness of $h(\mathrm{sd}_A(\sigma_n), x)$, the $h$-polynomial of the antiprism triangulation of the $(n-1)$-simplex.

**Theorem 0.1.** *The polynomial $h(\mathrm{sd}_A(\sigma_n), x)$ is real-rooted and has a nonnegative, real-rooted and interlacing symmetric decomposition with respect to $n-1$ for every positive integer $n$.*

We start by introducing the antiprism construction and its effect on the $h$-vector. Then using the antiprism construction, we define the antiprism triangulation in the second section of the chapter, and also give several other ways of defining it. In the third section we consider the antiprism triangulation of the simplex, and we prove combinatorial interpretations of its $h$-vector and Theorem 0.1 by using the recurrence relations for the $h$-polynomial of $\mathrm{sd}_A(\sigma_n)$ and the concept of interlacing sequences for polynomials. Furthermore, for a general simplicial complex with non-negative $h$-vector, we conjecture the real-rootedness of its $h$-polynomial and reduce it to an interlacing property of two related polynomials. We also provide several combinatorial interpretations of the local $h$-vector of the antiprism triangulation of the simplex. We close this section by generalizing the combinatorial interpreta-tions for the $h$-vector of the antiprism triangulation of the simplex and prove several combinatorial interpretations of the $h$-vector transformation for the antiprism triangulation of simplicial complexes.

The topic of the fourth chapter of this book are Lefschetz properties. We start by stating the classical $g$-theorem, defining the Lefschetz property and reviewing some results on classes of simplicial complexes with this property. In the second section of the chapter, we obtain the second main result, showing that the antiprism triangulation of a shellable simplicial complex has the almost strong Lefschetz property.

**Theorem 0.2.** *The antiprism triangulation $\mathrm{sd}_A(\Delta)$ has the almost strong Lefschetz property over $\mathbb{R}$ for every shellable simplicial complex $\Delta$.*

*Moreover, for every $(n-1)$-dimensional Cohen–Macaulay simplicial complex $\Delta$, the $h$-vector of $\mathrm{sd}_A(\Delta)$ is unimodal, with the peak being at position $n/2$, if $n$ is even, and at $(n-1)/2$ or $(n+1)/2$, if $n$ is odd.*

By introducing the *Strong Link Condition*, the Lefschetz property is shown first for the antiprism triangulation of the simplex, and then for a general shellable complex using the same proof as for the barycentric subdivision in [KN09]. Furthermore, it can be shown along the lines that the antiprism triangulation of the simplex is also strongly edge decomposable.

In the last chapter we shortly comment on some possible further questions regarding the antiprism triangulation.

# 1 Preliminaries

## 1.1 Simplicial complexes

In this section we introduce basic definitions and properties of simplicial complexes. In the first part we state basic definitions, such as the $f$- and $h$-vector which are combinatorial invariants associated to a simplicial complex.

In the second section we define the Stanley-Reisner ring, the algebraic object associated to a simplicial complex. Many of its algebraic invariants can be related to combinatorial invariants of the simplicial complex.

The third chapter reviews some different classes of simplicial complexes and their basic properties, as well as relations between some of them.

The last chapter contains the Kruskal-Katona inequalities, that characterize the $f$-vectors of simplicial complexes, and the Frankl-Füredi-Kalai inequalities characterizing the $f$-vectors of balanced simplicial complexes.

A great overview on simplicial complexes including the most important definitions and results can be found in [Sta96] and [BH93], a detailed description of the inequalities characterizing $f$-vectors can also be found in [Pet15].

### 1.1.1 Basic definitions

**Definition 1.1.** A *simplicial complex* $\Delta$ on vertex set $V$ is a set of subsets of $V$ such that the following properties hold:

(i) $\emptyset \in \Delta$,

(ii) if $F \in \Delta$ and $G \subseteq F$, then $G \in \Delta$.

The elements of a simplicial complex are called *faces*, and the *dimension* of a face $F \in \Delta$ is defined to be the cardinality of the face $F$ minus one, i.e., $\dim(F) = |F| - 1$. Now the *dimension* of the simplicial complex $\Delta$ is defined to be the maximal dimension of any of its faces, so

$$\dim(\Delta) = \max\{\dim(F) : F \in \Delta\}.$$

The 0-dimensional faces of a simplicial complex are called *vertices*, and 1-dimensional faces are called *edges*. Faces, which are maximal inclusion-wise, are called *facets*. Clearly, a simplicial complex is already completely determined by the set of its facets. A simplicial complex $\Delta$ is called *pure*, if all of its facets have the same dimension. A set $F \subseteq V$ with $F \notin \Delta$ is called a *non-face* of $\Delta$.

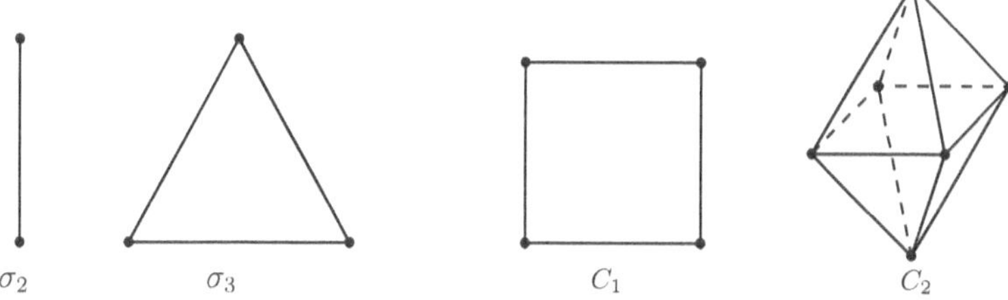

$\sigma_2$         $\sigma_3$         $C_1$         $C_2$

Figure 1.1: The 1- and 2-simplex, and the boundary complex of the 2- and 3-dimensional cross-polytope

We describe two important examples of simplicial complexes, which will serve as running examples in this section.

**Example 1.2.** We denote by $\sigma_n$ the $(n-1)$-dimensional simplicial complex which consists of all subsets of $[n]$, also called $(n-1)$-*simplex*.
Let $C_{n-1}$ be the boundary complex of the $n$-dimensional cross-polytope. Then $C_{n-1}$ is the $(n-1)$-dimensional simplicial complex on vertex set $V = \{v_1, \ldots, v_n, u_1, \ldots, u_n\}$, where the faces are subsets of $V$ which contain at most one element of $\{v_i, u_i\}$ for all $1 \le i \le n$.
In Figure 1.1 we see the 1- and 2-dimensional examples of these complexes.

In the following we consider some subsets of simplicial complexes and constructions on simplicial complexes, that yield new simplicial complexes from given ones.

**Definition 1.3.** Let $\Delta$ be a simplicial complex. We define the *link* of a face $F \in \Delta$ to be

$$\mathrm{lk}_\Delta(F) = \{G \in \Delta : G \cup F \in \Delta, G \cap F = \emptyset\},$$

and the *star* of the face $F \in \Delta$ to be

$$\mathrm{st}_\Delta(F) = \{G \in \Delta : G \cup F \in \Delta\}.$$

The link and the star of a face $F \in \Delta$ are again simplicial complexes, which describe the local structure around a face of the complex. For a pure $(n-1)$-dimensional simplicial complex $\Delta$ and a $k$-dimensional face $F \in \Delta$, $\mathrm{lk}_\Delta(F)$ is an $(n-2-k)$-dimensional simplicial complex.
Figure 1.2 shows an example of the link and the star of a face in a simplicial complex.

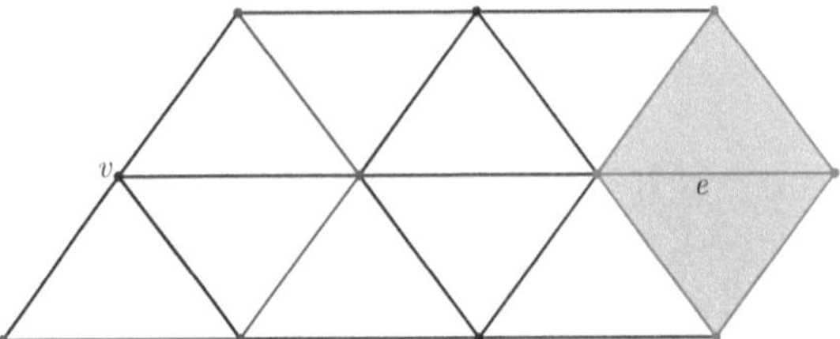

Figure 1.2: The link of the vertex $v$ (blue) and the star of the edge $e$ (red)

**Definition 1.4.** Let $\Delta$ be a simplicial complex on vertex set $V$. For a subset $W \subseteq V$, the *induced subcomplex* $\Delta[W]$ is the simplicial complex consisting of all faces in $\Delta$ whose vertices lie in $W$.

We define two more special subsets of simplicial complexes, the boundary and the interior.

**Definition 1.5.** Let $\Delta$ be a simplicial complex. The *boundary complex* of $\Delta$ is defined as

$$\partial(\Delta) \;=\; \langle F \in \Delta \;:\; F \subseteq G \text{ for a unique facet } G \in \Delta \rangle \cup \{\emptyset\}.$$

The set $\Delta^\circ = \Delta \setminus \partial(\Delta)$ consists of the *interior faces* of $\Delta$.

If $\partial(\Delta) \neq \emptyset$, the interior of the simplicial complex $\Delta$ is not a simplicial complex itself.

Given two simplicial complexes, there is the following construction to obtain a new simplicial complex.

**Definition 1.6.** Let $\Delta_1, \Delta_2$ be two simplicial complexes on disjoint vertex sets. The *(simplicial) join* of $\Delta_1$ and $\Delta_2$ is defined by

$$\Delta_1 * \Delta_2 = \{F_1 \cup F_2 : F_1 \in \Delta_1, F_2 \in \Delta_2\}.$$

If one of the two complexes consists of a single vertex $u$, e.g., $\Delta_1 = \{\{u\}, \emptyset\}$, then we call the join of these two complexes the *cone* of $\Delta_2$ with *apex* $u$, and it is denoted by $u * \Delta_2$.

**Example 1.7.** The simplicial complex $C_{n-1}$ is the join of the boundary of the 1-simplex and the complex $C_{n-2}$, i.e., $C_{n-1} = C_{n-2} * \partial(\sigma_2)$.

The $(n-1)$-dimensional simplex $\sigma_n$ can be described as the cone of the $(n-2)$-dimensional simplex $\sigma_{n-1}$.

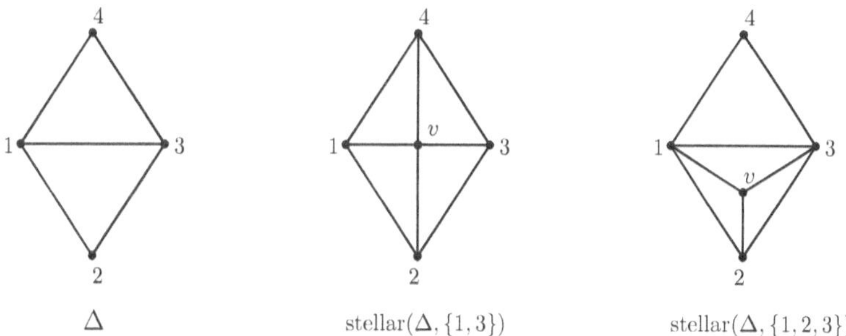

Figure 1.3: Stellar subdivisions of $\Delta$

**Definition 1.8.** For a simplicial complex $\Delta$ on vertex set $V$, the *stellar subdivision* on a face $F \in \Delta$ yields a new simplicial complex stellar$(\Delta, F)$ by replacing $\mathrm{st}_\Delta(F)$ by the join of $\mathrm{lk}_\Delta(F)$ and the cone over $\partial(2^F)$, i.e.,

$$\mathrm{stellar}(\Delta, F) = (\Delta \smallsetminus (F * \mathrm{lk}_\Delta(F))) \cup (\{v_F\} * \partial(F) * \mathrm{lk}_\Delta(F)),$$

where $v_F$ is a new vertex not in $V$.

Figure 1.3 shows examples of stellar subdivisions on different faces of a simplicial complex $\Delta$.

An important and much studied class of simplicial complexes are simplicial spheres.

**Definition 1.9.** We call a simplicial complex $\Delta$ *simplicial sphere*, if $\Delta$ triangulates a $(\dim(\Delta))$-dimensional sphere, meaning that the geometric realization of $\Delta$ is homeomorphic to a $(\dim(\Delta))$-dimensional sphere. More generally, a pure simplicial complex $\Delta$ is called a *homology sphere*, if for every face $F \in \Delta$, $\mathrm{lk}_\Delta(F)$ has the same homology as a $\dim(\mathrm{lk}_\Delta(F))$-sphere.

Observe that if $\Delta$ is a sphere, it holds that $\partial(\Delta) = \emptyset$.

One of the main problems in the combinatorics of simplicial complexes is the enumeration of their faces and bounds for the number of $k$-dimensional faces of a complex on a fixed vertex set $V$, for different classes of complexes.

For example, if $\Delta$ is a simplicial complex on vertex set $[n]$, a trivial upper bound for the number of $k$-dimensional faces is $\binom{n}{k+1}$, since the $k$-dimensional faces are $(k + 1)$-element subsets of the vertex set. One is now interested in (sharp) bounds for these numbers, and how the bounds might differ for different classes of simplicial complexes.

To study these problems, one introduces the so-called *face-vector* or $f$-vector, which collects the number of faces of the simplicial complex.

**Definition 1.10.** Let $\Delta$ be an $(n-1)$-dimensional simplicial complex. The *f-vector* $f(\Delta) = (f_{-1}(\Delta), f_0(\Delta), \dots, f_{n-1}(\Delta))$ is defined by $f_i(\Delta) := |\{F \in \Delta : \dim(F) = i\}|$ for $-1 \le i \le n-1$. The *f-polynomial* of $\Delta$ is defined as

$$f(\Delta, x) = \sum_{i=0}^{n} f_{i-1}(\Delta)x^i.$$

For example for a simplicial complex $\Delta \ne \emptyset$, we have that $\emptyset \in \Delta$ is the only $(-1)$-dimensional face, hence $f_{-1}(\Delta) = 1$.

**Example 1.11.** For the $(n-1)$-simplex we get $f_i(\sigma_n) = \binom{n}{i+1}$, since the $i$-dimensional faces are the $(i+1)$-element subsets of $[n]$.
The $i$-faces of $C_{n-1}$ can be counted in the following way: Choose an $(i+1)$-element subset of $[n]$, and then for each element $j$ of this subset choose either the vertex $v_j$ or $u_j$, so one gets $f_i(C_{n-1}) = 2^{i+1}\binom{n}{i+1}$.

Another important invariant, which is directly related to the $f$-vector and often better to work with, is the $h$-vector.

**Definition 1.12.** Let $\Delta$ be an $(n-1)$-dimensional simplicial complex. The *h-vector* of $\Delta$, denoted by $h(\Delta) = (h_0(\Delta), \dots, h_n(\Delta))$, is defined by

$$\sum_{i=0}^{n} h_i(\Delta)x^i = \sum_{i=0}^{n} f_{i-1}(\Delta)x^i(1-x)^{n-i} \tag{1.1}$$

$$= \sum_{F \in \Delta} x^{|F|}(1-x)^{n-|F|}. \tag{1.2}$$

The polynomial $h(\Delta, x) = \sum_{i=0}^{n} h_i(\Delta)x^i$ is called the *h-polynomial* of $\Delta$.
Similarly, $h^\circ(\Delta, x)$ is defined by the sum in (1.2) in which $\Delta$ has been replaced by $\Delta^\circ$.

So the $h$-polynomial and the $f$-polynomial of a simplicial complex are related by

$$h(\Delta, x) = (1-x)^n f\left(\Delta, \frac{x}{1-x}\right).$$

The $f$-vector and the $h$-vector of a simplicial complex can be obtained from one another in the following way.

**Proposition 1.13.** *Let $\Delta$ be an $(n-1)$-dimensional simplicial complex. Then*

$$h_j(\Delta) = \sum_{i=0}^{j}(-1)^{j-i}\binom{n-i}{j-i}f_{i-1}(\Delta) \tag{1.3}$$

*and*

$$f_{j-1}(\Delta) = \sum_{i=0}^{j}\binom{n-i}{j-i}h_i(\Delta). \tag{1.4}$$

**Example 1.14.** The $h$-vector of the $(n-1)$-simplex is $h(\sigma_n) = (1, 0, \ldots, 0)$.
The boundary of the $(n-1)$-simplex $\partial(\sigma_n)$ has the $h$-vector $h(\partial(\sigma_n)) = (1, 1, \ldots, 1)$.
For the boundary of the $n$-dimensional cross-polytope, we get $h_i(C_{n-1}) = \binom{n}{i}$.

Although in these two examples the entries of the $h$-vector are nonnegative, this
is not the case in general. Consider for example the complex $\Delta$ on vertex set $V = [5]$
with facets $\{1, 2, 3\}$ and $\{3, 4, 5\}$. The $f$-vector of $\Delta$ is $f(\Delta) = (1, 5, 6, 2)$ and its
$h$-vector is $h(\Delta) = (1, 2, -1, 0)$.
By Equation (1.4) we see, that bounds for the $h$-vector also give bounds for the
$f$-vector, since it is a nonnegative linear combination of the entries of the $h$-vector.

The following well known statement is a special case of [Sta87, Lemma 6.2], which
relates the $h$-polynomial of a simplicial complex triangulating a ball with the $h$-
polynomial of the interior of the complex.

**Proposition 1.15.** [Sta87] *We have $x^n h(\Delta, 1/x) = h^\circ(\Delta, x)$ for every triangulation
$\Delta$ of an $(n-1)$-dimensional ball.*

### 1.1.2 The Stanley-Reisner ring

Up to now, simplicial complexes are completely combinatorial objects, but to a
simplicial complex one can also relate an algebraic object, the so-called *Stanley-
Reisner ring*.

**Definition 1.16.** Let $\Delta$ be a simplicial complex on vertex set $[n]$ and let $\mathbb{F}$ be a
field. The *Stanley-Reisner ideal* $I_\Delta$ of $\Delta$ is the ideal $I_\Delta \subset \mathbb{F}[x_1, \ldots, x_n]$ generated
by the monomials $x_F = \prod_{i \in F} x_i$ for all $F \subseteq [n]$ with $F \notin \Delta$, i.e., the non-faces of $\Delta$.
   The *Stanley-Reisner ring* $\mathbb{F}[\Delta]$ is the standard graded $\mathbb{F}$-algebra

$$\mathbb{F}[\Delta] = \mathbb{F}[x_1, \ldots, x_n]/I_\Delta.$$

The Stanley-Reisner ideals are squarefree monomial ideals, and their definition
gives a one-to-one correspondence between simplicial complexes and squarefree monomial ideals.

**Example 1.17.** The Stanley-Reisner ring of the $(n-1)$-simplex $\sigma_n$ is $\mathbb{F}[\sigma_n] =
\mathbb{F}[x_1, \ldots, x_n]$, i.e., the polynomial ring in $n$ variables, since $\sigma_n$ does not have any
non-face.
For the vertex set $V = \{v_1, \ldots, v_n, u_1, \ldots, u_n\}$ of $C_{n-1}$, associate the variables $x_i$
to the vertex $v_i$ and the variable $x_{n+i}$ to the vertex $u_i$ for all $1 \leq i \leq n$. The
minimal non-faces of $C_{n-1}$ are $\{v_i, u_i\}$ for all $1 \leq i \leq n$. So the Stanley-Reisner
ideal of $C_{n-1}$ is $I_{C_{n-1}} = (x_1 x_{n+1}, \ldots, x_n x_{2n})$, and as Stanley-Reisner ring we obtain
$\mathbb{F}[C_{n-1}] = \mathbb{F}[x_1, \ldots, x_{2n}]/I_{C_{n-1}}$.

Moving from the combinatorial side of simplicial complexes to the algebraic side of
Stanley-Reisner rings yields a translation from combinatorial invariants to algebraic
invariants and vice versa. One example of this translation is given by the following,
connecting the $h$-vector of a simplicial complex with the $\mathbb{Z}$-graded Hilbert series of
the associated Stanley-Reisner ring.

**Theorem 1.18.** [BH93, Theorem 5.1.7] *Let $\Delta$ be an $(n-1)$-dimensional simplicial complex. Then*

$$\mathrm{Hilb}(\mathbb{F}[\Delta], x) = \sum_{i=0}^{n} \frac{f_{i-1}(\Delta)x^i}{(1-x)^i}$$
$$= \frac{h_0(\Delta) + h_1(\Delta)x + \cdots + h_n(\Delta)x^n}{(1-x)^n}.$$

The second equality here follows directly from Equation (1.1) after multiplying the numerator and the denominator of the $i$-th term in the sum by $(1-x)^{n-i}$. So from the $h$-vector of a simplicial complex we directly obtain the Hilbert series of the Stanley-Reisner ring and vice versa. Furthermore, we also get a direct relation between the dimension of the simplicial complex and the Krull-dimension of the ring $\mathbb{F}[\Delta]$ by $\dim(\Delta) + 1 = \dim(\mathbb{F}[\Delta])$.

### 1.1.3 Classes of simplicial complexes

In this section we want to introduce several special classes of simplicial complexes, that will appear in later chapters.

#### 1.1.3.1 Flag complexes

**Definition 1.19.** A simplicial complex $\Delta$ is called *flag*, if all minimal non-faces of $\Delta$ are of cardinality 2.

**Example 1.20.** The simplicial complex $C_{n-1}$ is a flag complex, since the minimal non-faces are $\{v_i, u_i\}$ for all $1 \le i \le n$, while for example the boundary of the $(n-1)$-simplex for $n \ge 3$ is not a flag complex, since $[n] \notin \Delta$ is a minimal non-face.

**Definition 1.21.** Let $\Delta$ be a simplicial complex. The simplicial complex

$$\mathrm{skel}_i(\Delta) = \{F \in \Delta : \dim(F) \le i\}$$

is called the *$i$-skeleton* of $\Delta$.

Flag simplicial complexes are closely related to graphs. Let $\Delta$ be a flag simplicial complex on vertex set $V$ and $F = \{v_1, \ldots, v_k\} \subseteq V$. If $\{v_i, v_j\} \in \Delta$ for all $1 \le i \le j \le k$, then also $F \in \Delta$. Hence a flag simplicial complex is already completely determined by its 1-skeleton, i.e., the set of the faces of dimension less or equal to 1.

#### 1.1.3.2 Eulerian complexes

For a simplicial complex $\Delta$, one can define its *reduced Euler characteristic* $\widetilde{\chi}(\Delta)$ to be the alternating sum of its reduced Betti numbers, i.e., $\widetilde{\chi}(\Delta) = \sum_{i=0}^{n-1}(-1)^i\widetilde{\beta}_i(\Delta; \mathbb{F})$ for a field $\mathbb{F}$. This invariant is directly related to the $f$- and $h$-vector, since $\widetilde{\chi}(\Delta)$ can also be expressed by

$$\widetilde{\chi}(\Delta) = -f(\Delta, -1).$$

For an $(n-1)$-dimensional complex, by equation (1.3) it is easy to see that

$$h_n(\Delta) = \sum_{i=0}^{n} (-1)^{n-i} f_{i-1}(\Delta) = (-1)^n f(\Delta, -1) = (-1)^n \widetilde{\chi}(\Delta).$$

**Example 1.22.** For the $(n-1)$-simplex we have $f(\sigma_n, x) = (1+x)^n$, so for the reduced Euler characteristic we get $\widetilde{\chi}(\sigma_n) = 0$.

In the case of the boundary of the cross-polytope $C_{n-1}$ we have seen before that $h_n(C_{n-1}) = \binom{n}{n} = 1$, so we get $\widetilde{\chi}(C_{n-1}) = (-1)^n h_n(C_{n-1}) = (-1)^n$.

**Definition 1.23.** A simplicial complex $\Delta$ is an *Eulerian complex*, if

$$\widetilde{\chi}(\mathrm{lk}_\Delta(F)) = (-1)^{\dim(\mathrm{lk}_\Delta(F))} \tag{1.5}$$

for every face $F \in \Delta$.

We say $\Delta$ is *semi-Eulerian*, if (1.5) holds for all non-empty faces $F \in \Delta$.

If $\Delta$ is a pure $(n-1)$-dimensional (semi-)Eulerian complex, we get $\widetilde{\chi}(\mathrm{lk}_\Delta(F)) = (-1)^{n-1-(\dim(F)+1)}$. Using this fact and a counting argument for the $k$-faces of $\Delta$, one gets that

$$(-1)^{n-1} f_k(\Delta) = \sum_{k \le i \le n} (-1)^i f_i \binom{i}{k}.$$

Now using the transformation from the $f$- to the $h$-vector, one obtains the Dehn-Sommerville equations for Eulerian complexes, first proved by Klee [Kle64].

**Theorem 1.24.** *Let $\Delta$ be a pure semi-Eulerian complex of dimension $n-1$. Then*

$$h_{n-k}(\Delta) - h_k(\Delta) = (-1)^k \binom{n}{k} (\widetilde{\chi}(\Delta) - (-1)^n)$$

*for all $0 \le k \le n$.*
*In particular, if $\Delta$ is Eulerian, then $h_{n-k}(\Delta) = h_k(\Delta)$ for $0 \le k \le n$.*

This theorem shows in particular, that the $h$-vectors of triangulations of spheres are symmetric. Two instances of this fact were already seen in Example 1.14, with $h_k(\partial(\sigma_n)) = 1 = h_{n-1-k}(\partial(\sigma_n))$ and $h_{n-k}(C_{n-1}) = \binom{n}{n-k} = \binom{n}{k} = h_k(C_{n-1})$.

### 1.1.3.3 Balanced complexes

**Definition 1.25.** A simplicial complex $\Delta$ is called *d-colorable*, if there is a partition of its vertex set $V = \bigcup_{i=1}^{d} V_i$ such that $|F \cap V_i| \le 1$ for each $F \in \Delta$ and $1 \le i \le d$.

Furthermore, an $(n-1)$-dimensional simplicial complex $\Delta$ is called *balanced*, if it is $n$-colorable.

Since for a facet $F \in \Delta$ all its vertices have to be in different classes, we immediately see that $d \geq n$ has to hold for an $(n-1)$-dimensional simplicial complex. Hence a simplicial complex is balanced, if it suffices to partition the vertex set into the smallest possible number of classes.

One can think of the vertices being colored with the colors $\{1, \ldots, n\}$ such that in each face $F \in \Delta$, all vertices of $F$ are of different colors.

This means the balanced complexes are the $(n-1)$-dimensional complexes $\Delta$, where the 1-skeleton of $\Delta$ is $n$-colorable in the graph-theoretical sense.

**Example 1.26.** The $(n-1)$-simplex is a balanced complex. The boundary of the $(n-1)$-simplex is not balanced, since every vertex is connected to every other vertex. Hence one needs $n$ colors to color the vertices such that no edge connects vertices of the same color and the 1-skeleton is $n$-colorable. Since the boundary of $\sigma_n$ is $(n-2)$-dimensional, it is not balanced.

Also $C_{n-1}$ is a balanced complex, since the partition $V = \cup_{i=1}^{n}\{v_i, u_i\}$ gives a partition of the vertex set in the right way.

### 1.1.3.4 Cohen–Macaulay complexes

**Definition 1.27.** Let $\Delta$ be a simplicial complex and $\mathbb{F}$ be an arbitrary field. $\Delta$ is called a *Cohen–Macaulay complex over* $\mathbb{F}$, or simply *Cohen–Macaulay over* $\mathbb{F}$, if its Stanley-Reisner ring is Cohen–Macaulay over $\mathbb{F}$, i.e., $\dim(\mathbb{F}[\Delta]) = \operatorname{depth}(\mathbb{F}[\Delta])$.

$\Delta$ is called *Cohen–Macaulay complex*, if it is Cohen–Macaulay over some field $\mathbb{F}$.

Since it can be quite difficult to decide whether a simplicial complex is Cohen–Macaulay by this definition, a helpful tool is *Reisner's criterion*.

**Theorem 1.28.** [BH93, Corollary 5.3.9] *Let $\Delta$ be a simplicial complex. Then the following are equivalent:*

*(i) $\Delta$ is Cohen–Macaulay over $\mathbb{F}$*

*(ii) $\widetilde{H}_i(\operatorname{lk}_\Delta(F); \mathbb{F}) = 0$ for all $F \in \Delta$ and $i < \dim(\operatorname{lk}_\Delta(F))$.*

By Reisner's criterion it follows, that every link of a Cohen–Macaulay complex is again a Cohen–Macaulay complex.

**Example 1.29.** Using Reisner's criterion we easily see, that the $(n-1)$-simplex $\sigma_n$ is Cohen–Macaulay, since for a $k$-face $F \in \sigma_n$, $\operatorname{lk}_{\sigma_n}(F)$ is an $(n-2-k)$-simplex, so the links are simplicial balls having homology zero in each dimension.

Also $C_{n-1}$ is a Cohen–Macaulay complex. Since $\operatorname{lk}_{C_{n-1}}(F)$ for $F \in C_{n-1}$ is again the boundary complex of a cross-polytope, the links are spheres. Hence they only have non-zero homology in the top dimension, and Reisner's criterion yields that $C_{n-1}$ is Cohen–Macaulay.

The complex $\Delta$ having facets $\{1,2,3\}$ and $\{3,4,5\}$ is not Cohen–Macaulay. We have that $\operatorname{lk}_\Delta(\{3\})$ is the complex having facets $\{1,2\}$ and $\{4,5\}$, i.e., it has two isolated edges as facets. It holds that $\tilde{H}_0(\mathrm{lk}_\Delta(\{3\}); \mathbb{F}) = \mathbb{F}$ since it has two connected components. This means $(ii)$ of Theorem 1.28 does not hold, hence $\Delta$ can not be Cohen–Macaulay by Reisner's criterion.

### 1.1.3.5 Shellable complexes

**Definition 1.30.** A simplicial complex $\Delta$ is called *shellable*, if $\Delta$ is a simplex or if the following equivalent conditions are satisfied:
There exists a linear ordering $F_1, \ldots, F_l$ of the facets of $\Delta$ such that

(i) $\langle F_i \rangle \cap \langle F_1, \ldots, F_{i-1} \rangle$ is generated by a non-empty set of maximal proper faces of $\langle F_i \rangle$, for all $2 \leq i \leq l$, where we define $\langle F_1, \ldots, F_{i-1} \rangle$ to be the simplicial complex whose faces are all subsets of $F_1, \ldots, F_{i-1}$.

(ii) The set $\{F : F \in \langle F_1, \ldots, F_i \rangle, F \notin \langle F_1, \ldots, F_{i-1} \rangle\}$ has a unique minimal element for all $2 \leq i \leq l$. This element is called the *restriction face* of $F_i$, denoted by $\mathrm{res}(F_i)$.

A linear order of the facets satisfying these equivalent conditions is called a *shelling* of $\Delta$.

Shellability is a stronger property than being Cohen–Macaulay, there is the following relation.

**Theorem 1.31.** [BH93, Theorem 5.1.13] *Let $\Delta$ be a shellable simplicial complex. Then $\Delta$ is Cohen–Macaulay over any field $\mathbb{F}$.*

This implication is strict, so it is not true, that all Cohen–Macaulay complexes are shellable, see the next example.

**Example 1.32.** The boundary of the $(n-1)$-simplex is a shellable complex, any ordering of its facets gives a shelling. The complex $C_{n-1}$ is shellable too, but not every ordering of the facets yields a shelling. For example the first two facets of the shelling can not be $\{u_1, u_2, \ldots, u_n\}$ and $\{v_1, \ldots, v_n\}$, since $\langle\{u_1, \ldots, u_n\}\rangle \cap \langle\{v_1, \ldots, v_n\}\rangle = \emptyset$, which does not fulfill condition (i) of Definition 1.30.
Since the complex $\Delta$ having facets $\{1, 2, 3\}$ and $\{3, 4, 5\}$ is not Cohen–Macaulay as seen in the previous section, it can not be shellable either.
An example showing that the implication in Theorem 1.31 is strict, i.e., there are Cohen–Macaulay complexes that are not shellable, are triangulations of the dunce hat. The dunce hat is a topological space, that can be thought of as a triangle whose edges are identified with one orientation reversed. It can be shown, that every triangulation of the dunce hat is Cohen–Macaulay over any field, but no triangulation is shellable (see for example [Sta96, Chapter III.2]).

Although the two classes of shellable and Cohen–Macaulay complexes do not coincide, they share the same set of $h$-vectors, meaning that every $h$-vector of a shellable complex is the $h$-vector of a Cohen–Macaulay complex and vice versa.

**Theorem 1.33.** [BH93, Theorem 5.1.5] *Let $h = (h_0, \ldots, h_n)$ be a sequence of integers the following are equivalent:*

*(i) $h$ is the $h$-vector of a shellable complex.*

*(ii) $h$ is the $h$-vector of a Cohen–Macaulay complex.*

For a shellable simplicial complex $\Delta$, there is a nice combinatorial interpretation of its $h$-vector $h(\Delta)$.

**Theorem 1.34.** [BH93, Corollary 5.1.14] *Let $\Delta$ be an $(n-1)$-dimensional simplicial complex with a shelling $F_1, \ldots, F_l$. Let $r_j$ be the number of facets of $\langle F_j \rangle \cap \langle F_1, \ldots, F_{j-1} \rangle$ for $2 \leq j \leq l$ and $r_1 = 0$. Then*

$$h_i(\Delta) = |\{j : r_j = i\}|$$

*for all $0 \leq i \leq n$.*
*In particular, up to their order, the numbers $r_j$ do not depend on the shelling.*

One can also state this theorem in terms of the second condition in Definition 1.13, since $r_j = \dim(\mathrm{res}(F_j)) + 1$ by the definition of the restriction faces.

In particular, Theorem 1.34 together with Theorem 1.33 shows that the $h$-vectors of shellable complexes and of Cohen–Macaulay complexes are nonnegative.

### 1.1.4 Characterization of $f$-vectors

In this section we review the relations between the entries of the $f$-vector, which characterize the $f$-vectors of arbitrary simplicial complexes on the one hand and balanced complexes on the other hand.

#### 1.1.4.1 Kruskal-Katona inequalities

First we want to give the characterization of face vectors of any simplicial complex without any further restrictions or properties. To do so, we need the following lemma:

**Lemma 1.35.** [BH93, Lemma 4.2.6] *Let $a, i > 0$ be two integers. Then there exist unique numbers $m_i > m_{i-1} > \cdots > m_l \geq l \geq 1$, such that*

$$a = \binom{m_i}{i} + \binom{m_{i-1}}{i-1} + \cdots + \binom{m_l}{l}.$$

*This is called the $i$-binomial representation of $a$.*

Now given two integers $a, i > 0$ and the representation as in the previous Lemma, we define the number $a^{(i)}$ to be

$$a^{(i)} = \binom{m_i}{i+1} + \binom{m_{i-1}}{i} + \cdots + \binom{m_l}{l+1}.$$

With this definition we can characterize the $f$-vectors of simplicial complexes, shown by Kruskal [Kru63] and Katona [Kat68].

**Theorem 1.36.** *The following are equivalent:*

*(i) The vector $(1, f_0, \ldots, f_{n-1}) \in \mathbb{Z}^{n+1}$ is the $f$-vector of an $(n-1)$-dimensional simplicial complex.*

*(ii) $0 < f_{i+1} \leq f_i^{(i+1)}$ for all $0 \leq i \leq n-2$.*

To construct a simplicial complex with an $f$-vector given as in $(ii)$, one associates to this vector a so-called *compressed complex* $C$, where the $i$-dimensional faces of $C$ are the first $f_i$-many $(i+1)$-element subsets of $\mathbb{N}$ ordered by the so-called *co-lex order*, defined as follows. Let $A$ and $B$ be two $k$-element sets of $\mathbb{N}$, and let $A \triangle B$ be the symmetric difference of $A$ and $B$. Now we say $A \prec B$ in the co-lex ordering if and only if $\max(A \triangle B) \in B$. So for instance, for $k = 3$ the smallest sets according to this order are $123 \prec 124 \prec 134 \prec 234 \prec 125 \prec 135 \prec 145 \prec 235 \prec 245 \prec \ldots$. Now $C$ has the wanted $f$-vector and one can show that it is actually a simplicial complex.

On the other hand, for a given simplicial complex $\Delta$, on can transform it via *compression* operations on the faces of $\Delta$ into another simplicial complex with the same $f$-vector. For this complex, one can show the desired inequalities.

### 1.1.4.2 Frankl-Füredi-Kalai inequalities

The Kruskal-Katona inequalities give the characterization for all possible $f$-vectors of arbitrary simplicial complexes. Nevertheless, for more special and restricted classes of simplicial complexes one might expect stronger inequalities to hold for the $f$-vectors of these simplicial complexes.

Exactly this was shown by Frankl, Füredi and Kalai [FFK88] by providing inequalities that are stronger than the Kruskal-Katona inequalities, and which characterize the $f$-vectors of $d$-colorable simplicial complexes analog to Theorem 1.36. For showing these stronger inequalities, they introduce a colored version of the theory used for the proof of Theorem 1.36. Define a *$d$-colored $k$-set* of $\mathbb{N}$ to be a $k$-element set $S$ such that no two elements of $S$ are congruent modulo $d$, and denote by $\binom{\mathbb{N}}{k}_d$ the set of all $d$-colored $k$-sets. The number of $d$-colored $k$-sets with maximal element smaller than $n$ is denoted by

$$\binom{n}{k}_d = |\{S \in \binom{\mathbb{N}}{k}_d : \max(S) < n\}|.$$

One can think of this as colored version of the binomial coefficients, since $\binom{n}{k}_d = \binom{n}{k}$ for $n \leq d$.

Now one gets the following result [FFK88, Lemma 1.1] analog to Lemma 1.35.

**Lemma 1.37.** [FFK88] *Let $a, d \geq i$ be positive integers. Then there exist unique numbers $m_i > m_{i-1} > \cdots > m_{i-l} \geq i - l \geq 1$, such that*

$$a = \binom{m_i}{i}_d + \binom{m_{i-1}}{i-1}_{d-1} + \cdots + \binom{m_{i-l}}{i-l}_{d-l}.$$

Define $a_d^{(i)}$ to be

$$a_d^{(i)} = \binom{m_i}{i+1}_d + \binom{m_{i-1}}{i}_{d-1} + \cdots + \binom{m_{i-l}}{i-l+1}_{d-l}.$$

Using this representation, one can prove the result [FFK88, Theorem 1.2], yielding an analogue to Theorem 1.36.

**Theorem 1.38.** *The following are equivalent:*

*(i) The vector $(1, f_0, \ldots, f_{n-1}) \in \mathbb{Z}^{n+1}$ is the $f$-vector of a $d$-colorable $(n-1)$-dimensional simplicial complex.*

*(ii) $0 \leq f_{i+1} \leq (f_i)_d^{(i+1)}$ for all $0 \leq i \leq n - 2$.*

Similar as in the proof of the Kruskal-Katona inequalities, one constructs a $d$-colorable simplicial complex, by taking the first $d$-colored $k$-sets in the ordering one obtains by restricting the co-lex order to the $d$-colored $k$-sets.

Furthermore, Frohmader [Fro08, Theorem 1.1] could show, that every $f$-vector of a flag simplicial complex is also the $f$-vector of a balanced simplicial complex.

**Theorem 1.39.** [Fro08] *For any flag simplicial complex $\Delta$ there is a balanced simplicial complex $\Delta'$, such that $f(\Delta) = f(\Delta')$.*

The converse of this statement is not true, as the example of a vector satisfying the FFK-inequalites that can not be the $f$-vector of a flag somplex described in [Fro08, Section 3] shows.

## 1.2 Statistics on permutations

In this section we want to shortly review one of the most studied numbers in enumerative combinatorics, the Eulerian numbers. We define the Eulerian numbers by several statistics on the set of permutations, and state some basic properties, such as recursive and explicit descriptions. Furthermore we define the Eulerian polynomial and the closely related derangement polynomial. A detailed overview on the Eulerian numbers and related topics can be found in [Pet15], as well as in [Sta12].

In the second part of the section the generalization of colored permutations is introduced, and the colored versions of the Eulerian polynomial and derangement polynomial are defined.

### 1.2.1 Eulerian numbers

**Definition 1.40.** For a positive integer $n$, let the *symmetric group* $\mathfrak{S}_n$ be the set of all permutations on $[n]$, i.e., all bijections $w : [n] \to [n]$. We will write elements of $\mathfrak{S}_n$ as $w = w(1)w(2)\ldots w(n)$. We define the following statistics on the set of permutations: For $w \in \mathfrak{S}_n$ we call

(i) $\operatorname{Asc}(w) = \{i \in [n-1] : w(i) < w(i+1)\}$ the set of *ascents*, and its cardinality is denoted by $\operatorname{asc}(w)$,

(ii) $\operatorname{Des}(w) = \{i \in [n-1] : w(i) > w(i+1)\}$ the set of *descents*, and its cardinality is denoted by $\operatorname{des}(w)$,

(iii) $\operatorname{Exc}(w) = \{i \in [n-1] : w(i) > i\}$ the set of *excedances* and its cardinality is denoted by $\operatorname{exc}(w)$,

(iv) $\operatorname{Wexc}(w) = \{i \in [n] : w(i) \geq i\}$ the set of *weak excedances* and its cardinality is denoted by $\operatorname{wexc}(w)$.

**Example 1.41.** Let $w = 351726984 \in S_9$. Then we have $\operatorname{Asc}(w) = \{1,3,5,6\}$, $\operatorname{Des}(w) = \{2,4,7,8\}$, $\operatorname{Exc}(w) = \{1,2,4,7\}$ and $\operatorname{Wexc}(w) = \{1,2,4,6,7,8\}$.

As can be observed in the example, it is clear that $\operatorname{Asc}(w) \cup \operatorname{Des}(w) = [n-1]$. Furthermore, there is the bijection on $\mathfrak{S}_n$ mapping $w \in \mathfrak{S}_n$ to $\tilde{w} \in \mathfrak{S}_n$ with $\tilde{w}(i) = n + 1 - w(i)$ for $1 \leq i \leq n$, such that $\operatorname{Asc}(w) = \operatorname{Des}(\tilde{w})$ and $\operatorname{Des}(w) = \operatorname{Asc}(\tilde{w})$.

**Definition 1.42.** For integers $n \geq 1$ and $1 \leq k \leq n - 1$ we define the *Eulerian numbers* $A(n, k)$ to be the number of permutations $w \in \mathfrak{S}_n$ with $k$ descents, i.e.,

$$A(n, k) = |\{w \in \mathfrak{S}_n : \operatorname{des}(w) = k\}|.$$

In the literature, the Eulerian numbers are also often denoted by $\left\langle {n \atop k} \right\rangle$.

**Example 1.43.** For $n = 3$ we get the following:

1. $A(3, 0) = 1$, since the only permutation without any descent is $123$.

2. $A(3, 1) = 4$, since the permutations having one descent are $132, 213, 231, 312$.

3. $A(3, 2) = 1$, since the only permutation with two descents is $321$.

In Table 1 there are more examples of $A(n, k)$ for small $n$.

| $n \backslash k$ | 0 | 1 | 2 | 3 | 4 | 5 |
|---|---|---|---|---|---|---|
| 1 | 1 | | | | | |
| 2 | 1 | 1 | | | | |
| 3 | 1 | 4 | 1 | | | |
| 4 | 1 | 11 | 11 | 1 | | |
| 5 | 1 | 26 | 66 | 26 | 1 | |
| 6 | 1 | 57 | 302 | 302 | 57 | 1 |

Table 1.1: Eulerian numbers for $1 \leq n \leq 6$

Since the only permutation without any descents is $w = 123\ldots n-1n$, and the only permutation with $n-1$ descents is $w = nn-1\ldots 21$, it is clear that $A(n,0) = A(n,n-1) = 1$. Similarly, using the bijection described above, we see that this symmetry also holds in general, so we have $A(n,k) = A(n,n-k-1)$ for $1 \leq k \leq n-1$.

Furthermore, also by this bijection, we see that counting the permutations of $\mathfrak{S}_n$ with $k$ ascents yields the Eulerian numbers too, we have

$$A(n,k) = \{w \in \mathfrak{S}_n : \mathrm{asc}(w) = k\}.$$

There are several other characterizations of the Eulerian numbers, including

$$A(n,k) = \{w \in \mathfrak{S}_n : \mathrm{exc}(w) = k\} = \{w \in \mathfrak{S}_n : \mathrm{wexc}(w) = k-1\},$$

where the first equality can be obtained via a bijection mapping descents to excedances, called "transformation fondamentale" by Foata and Schützenberger [FS70].

The Eulerian numbers admit a nice recurrence relation, obtained in the following way: If a permutation $w \in \mathfrak{S}_n$ is given with $k$ descents, deleting the value $n$ from that permutation will result in a permutation $w' \in \mathfrak{S}_{n-1}$ with either $k$ or $k-1$ descents. On the other hand, consider an element $w \in \mathfrak{S}_{n-1}$ with $k-1$ descents. One obtains an element $w' \in \mathfrak{S}_n$ with $k$ descents by inserting the element $n$ in $w$ at a position, where $w$ has an ascent, or at the beginning. There are $n-k$ of these positions. Now consider an element $w \in \mathfrak{S}_{n-1}$ with $k$ descents. Here one has to insert the element $n$ at a position of a descent or at the end of $w$ to obtain an element $w' \in \mathfrak{S}_n$ with $k$ descents, and there are $(k+1)$ of these positions. This results in the recursion

$$A(n,k) = (n-k)A(n-1,k-1) + (k+1)A(n-1,k). \tag{1.6}$$

An explicit expression for $A(n,k)$ is given by

$$A(n,k) = \sum_{i=0}^{k}(-1)^i \binom{n+1}{i}(k+1-i)^n.$$

Let $S(n,k)$ be the Stirling number of the second kind. Then there is the following relation between these numbers and the Eulerian numbers [Cha02]:

$$A(n,k) = \sum_{i=0}^{k}(-1)^{k-i}\binom{n-i}{k-i}i!S(n,i). \tag{1.7}$$

The Eulerian polynomial is defined as the generating function of the Eulerian numbers for a fixed $n$.

**Definition 1.44.** The *n-th Eulerian polynomial* $A_n(x)$ is the polynomial

$$A_n(x) = \sum_{k=0}^{n-1} A(n,k)x^k = \sum_{w \in \mathfrak{S}_n} x^{\mathrm{des}(w)}.$$

**Example 1.45.** From Table 1 we obtain that the first few Eulerian polynomials are given by

$$A_2(x) = 1 + x,$$
$$A_3(x) = 1 + 4x + x^2,$$
$$A_4(x) = 1 + 11x + 11x^2 + x^3.$$

Using the Recursion (1.6) yields the following recursive description of the Eulerian polynomials.

**Theorem 1.46.** *For $n \geq 0$,*

$$A_{n+1}(x) = (1 + nx)A_n(x) + x(1 - x)A_n'(x).$$

Closely related to these polynomials is the so-called *derangement polynomial*.

**Definition 1.47.** We say an index $i$, $1 \leq i \leq n$, is a fixed point of a permutation $w \in \mathfrak{S}_n$, if $w(i) = i$. We call a permutation a *derangement* if it does not have any fixed point, i.e., $w(i) \neq i$ for all $1 \leq i \leq n$. We denote by $D_n$ the set of *derangements* in $\mathfrak{S}_n$.

The *derangement polynomial* $d_n(x)$ is defined by

$$d_n(x) = \sum_{w \in D_n} x^{\mathrm{exc}(w)}.$$

**Example 1.48.** For $n = 3$, the set of derangements is $D_3 = \{231, 312\}$. The first derangement polynomials are

$$d_2(x) = x,$$
$$d_3(x) = x + x^2,$$
$$d_4(x) = x + 7x^2 + x^3.$$

By using the description of the Eulerian numbers in terms of excedances, one can describe the Eulerian polynomials and the derangement polynomials in terms of each other by

$$A_n(x) = \sum_{k=0}^{n} \binom{n}{k} d_k(x)$$

and

$$d_n(x) = \sum_{k=0}^{n} (-1)^{n-k} \binom{n}{k} A_k(x).$$

### 1.2.2 $r$-colored permutations

In this section we generalize the Eulerian polynomial and the derangement polynomial by considering now colored permutations.

**Definition 1.49.** Let $n$ and $r$ be positive integers. We call an element $(w, z) \in \mathbb{Z}_r \wr \mathfrak{S}_n$, given by $w \in \mathfrak{S}_n$ and $z \in \{0, 1, \ldots, r-1\}^n$, an $r$-*colored permutation*. We define some statistics on the colored permutations generalizing the statistics for permutations seen before. For $(w, z) \in \mathbb{Z}_r \wr \mathfrak{S}_n$ the index $i \in [n]$ is a *descent* if either $z_i > z_{i+1}$ or $z_i = z_{i+1}$ and $w_i > w_{i+1}$, where $w_{n+1} = n+1$ and $z_{n+1} = 0$. A *descending run* is a maximal string $\{a, a+1, \ldots, b\}$ of integers such that $i$ is a descent for every $a \leq i \leq b-1$. The index $i \in [n]$ is an *excedance* of $(w, z) \in \mathbb{Z}_r \wr \mathfrak{S}_n$ if $w(i) > i$ or $w(i) = i$ and $z_i > 0$. By $\mathrm{des}(w, z)$ and $\mathrm{exc}(w, z)$ we denote the number of descents of $(w, z)$ and the number of excedances of $(w, z)$ respectively. As in the case of permutations, these two statistics are distributed equally [Ste94, Theorem 15]. The *flag excedance* of $(w, z) \in \mathbb{Z}_r \wr \mathfrak{S}_n$ is defined as

$$\mathrm{fexc}(w, z) = r \cdot \mathrm{exc}_A(w, z) + \mathrm{csum}(w, z)$$

where $\mathrm{exc}_A(w, z)$ is the number of excedances with color zero and $\mathrm{csum}(w, z) = \sum_{i=0}^{n} z_i$ is the sum of all colors. Furthermore, we call an element $(w, z) \in \mathbb{Z}_r \wr \mathfrak{S}_n$ *balanced*, if $\mathrm{csum}(w, z)$ is divisible by $r$.

We define for $n \geq 1$ the $n$-*th Eulerian polynomial for $r$-colored permutations of* $[n]$ to be the polynomial

$$A_{n,r}(x) = \sum_{(w,z) \in \mathbb{Z}_r \wr \mathfrak{S}_n} x^{\mathrm{des}(w,z)} = \sum_{(w,z) \in \mathbb{Z}_r \wr \mathfrak{S}_n} x^{\mathrm{exc}(w,z)}.$$

We say $i \in [n]$ is a *fixed point of color zero* of an $r$-colored permutation $(w, z)$, if $w(i) = i$ and $z_i = 0$. A *derangement* is an $r$-colored permutation without any fixed point, and the set of all derangements is denoted by $D_n^r$. The $n$-*th derangement polynomial for the $r$-colored permutations on* $[n]$ is the polynomial

$$d_{n,r}(x) = \sum_{(w,z) \in D_n^r} x^{\mathrm{exc}(w,z)}.$$

By setting $r = 1$, we recover the classical $n$-th Eulerian polyomial and $n$-th derangement polynomial defined in the previous section.

## 1.3 Properties of polynomials

The aim of this section is to review some important and interesting properties of polynomials, in particular real-rootedness and $\gamma$-nonnegativity, and how these properties relate to each other. A great overview on the topics mentioned in this section can be found in a survey by Brändén [Brä15], as well as in [Fis08] on the topic of interlacing polynomials, and in the survey on $\gamma$-nonnegativity by Athanasiadis [Ath18].

**Definition 1.50.** Let $f(x) = \sum_{i=0}^{n} a_i x^i \in \mathbb{R}[x]$ be a polynomial. We call the polynomial

(i) *symmetric* with *center of symmetry* $\frac{n}{2}$, if $a_i = a_{n-i}$ for all $1 \le i \le n$,

(ii) *unimodal* if there is an index $0 \le j \le n$ such that

$$a_0 \le \cdots \le a_{j-1} \le a_j \ge a_{j+1} \ge \cdots \ge a_n,$$

(iii) *log-concave* if
$$a_j^2 \ge a_{j-1} a_{j+1}, \qquad \text{for all } 1 \le j \le n-1,$$

(iv) *real-rooted* if all its zeros are real, or it is a constant polynomial,

(v) *alternatingly increasing* if

$$a_0 \le a_n \le a_1 \le a_{n-1} \le \cdots \le a_{\lceil \frac{n}{2} \rceil}.$$

There are the following relations between the different properties mentioned in Definition 1.50. The first two statements can be found in [Brä15, Lemma 1.1], the third follows directly be the definition.

**Lemma 1.51.** [Brä15, Lemma 1.1] *Let* $f(x) = \sum_{i=0}^{n} a_i x^i$ *be a polynomial with nonnegative coefficients.*

*(i) If $f(x)$ is real-rooted, then it is log-concave.*

*(ii) If $f(x)$ is log-concave with positive coefficients, then it is unimodal.*

*(iii) If $f(x)$ is altenating increasing, then it is unimodal.*

**Example 1.52.**  (i) The polynomial $f(x) = 3x^4 + 13x^3 + 22x^2 + 13x + 3$ is symmetric with center of symmetry two and unimodal. Furthermore, it is also log-concave, since $13^2 \ge 3 \cdot 22$ and $22^2 \ge 13 \cdot 13$.

(ii) An important family of polynomials sharing these properties are the Eulerian polynomials. The symmetry of the Eulerian polynomial was already shown in the previous section. These polynomials are real-rooted, first shown by Frobenius (see for example in [Fro68]), hence also log-concave and unimodal. We will also see an idea of a proof of the real-rootedness later.
The more general polynomials $A_{n,r}$ were shown to be real-rooted by Steingrímsson [Ste94].

(iii) The derangement polynomials $d_n(x)$ are unimodal and symmetric, shown by Brenti [Bre90, Corollary 1]. It was proven by Zhang [Zha95], that these polynomials are also real-rooted.
For the more general polynomials $d_{n,r}(x)$, the real-rootedness was shown by Chow and Mansour [CM10, Theorem 5].

(iv) The converses of the statements in Lemma 1.51 are false in general, see [Brä15, Examples 1.1 and 1.2]. Examples of polynomials, which are log-concave but not real-rooted, are given by the $q$-factorial polynomials, the generating functions for the number of inversions over $\mathfrak{S}_n$, an example of such a polynomial is $f(x) = x^3 + 2x^2 + 2x + 1$. Examples, which are unimodal and symmetric but not log-concave, are the $q$-binomial coefficients, for which the polynomial $f(x) = x^4 + x^3 + 2x^2 + x + 1$ is an example.

For a symmetric polynomial there is another property that can be related to the already mentioned properties. Since a polynomial $f(x) = \sum_{i=0}^{n} a_i x^i$ is symmetric if and only if it can be written in the form

$$f(x) = \sum_{i=0}^{\lfloor \frac{n}{2} \rfloor} \gamma_i x^i (1+x)^{n-2i}, \tag{1.8}$$

one can consider the presentation of symmetric polynomials according to the basis $\{x^i(1+x)^{n-2i}\}_{i=0}^{\lfloor \frac{n}{2} \rfloor}$.

**Definition 1.53.** For a symmetric polynomial

$$f(x) = \sum_{i=0}^{\lfloor \frac{n}{2} \rfloor} \gamma_i x^i (1+x)^{n-2i}$$

we call $(\gamma_0, \ldots, \gamma_{\lfloor \frac{n}{2} \rfloor})$ the $\gamma$-*vector* of $f$. We say the polynomial $f$ is $\gamma$-*nonnegative* if its $\gamma$-vector is nonnegative.

The following fact can be found in [Brä06, Lemma 4.1] or in [Gal05, Remark 3.1.1].

**Lemma 1.54.** *Let $f$ be a symmetric polynomial. If $f$ is real-rooted, then it is $\gamma$-nonnegative.*

**Example 1.55.** (i) The polynomial $f(x) = 3x^4 + 13x^3 + 22x^2 + 13x + 3$ is $\gamma$-nonnegative, it can be written as $f(x) = 3(1+x)^4 + 1x(1+x)^2 + 2x^2$. The polynomial $f(x) = x^4 + 2x^3 + 6x^2 + 2x + 1$ is not $\gamma$-nonnegative, its $\gamma$-vector is $\gamma = (1, -2, 4)$.

(ii) The Eulerian polynomials and the derangement polynomials are $\gamma$-nonnegative, since they are real-rooted. There are also combinatorial interpretations of their $\gamma$-vectors, that we discuss in more detail in the next section.

(iii) Again, the converse of Lemma 1.54 is not true, for example the polynomial $f(x) = x^4 + 4x^3 + 7x^2 + 4x + 1$ has no real roots, and it has the nonnegative $\gamma$-vector $(1, 0, 1)$, since $f(x) = (1+x)^4 + x^2$.

But on the other hand, the $\gamma$-nonnegativity of a polynomial $f$ implies its unimodality since it can be written as the sum of symmetric and unimodal polynomials with the same center of symmetry.

Figure 1.4 summarizes the relation between the properties real-rootedness, log-concavity and $\gamma$-nonnegativity.

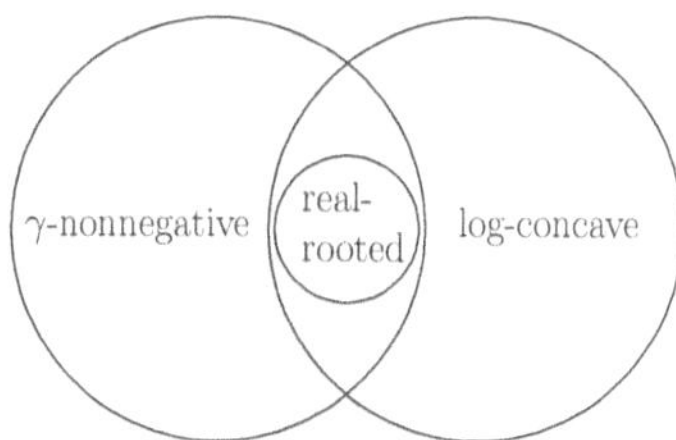


Figure 1.4: Relation between real-rootedness, log-concavity and $\gamma$-nonnegativity

Now we shortly sketch two of the many methods that have been used to prove real-rootedness of families of polynomials.

**Definition 1.56.** Let $f$ and $g$ be two real-rooted polynomials and let $\alpha_1 \geq \alpha_2 \geq \ldots$ and $\beta_1 \geq \beta_2 \geq \ldots$ be the roots of $f$ and $g$ respectively. We say that $f$ *interlaces* $g$ if
$$\beta_1 \geq \alpha_1 \geq \beta_2 \geq \alpha_2 \geq \ldots.$$
By convention we also say that the zero polynomial interlaces and is interlaced by every real-rooted polynomial.

We call a sequence of real-rooted polynomials $(f_1(x), f_2(x), \ldots, f_m(x))$ *interlacing*, if $f_i(x)$ interlaces $f_j(x)$ for all $1 \leq i < j \leq m$.

The next lemma collects some statements on interlacing polynomial which will be used later.

**Lemma 1.57.**    *1.* [Brä06, Lemma 2.3] *Let $f_1(x), \ldots, f_m(x) \in \mathbb{R}[x]$ be real-rooted polynomials. If $f_1(x)$ interlaces $f_m(x)$ and $f_i(x)$ interlaces $f_{i+1}(x)$ for all $i \in [n-1]$, then $(f_1(x), \ldots, f_m(x))$ is an interlacing sequence.*

*2.* [Fis08, Lemma 3.4] *Let $(f_1(x), \ldots, f_m(x))$ be an interlacing sequence of real-rooted polynomials with positive leading coefficients, then so is $(f_1(x) + \cdots + f_m(x), \ldots, f_{m-1}(x) + f_m(x), f_m(x))$.*

*3. Let $(f_1(x), \ldots, f_m(x))$ be an interlacing sequence of real-rooted polynomials with positive leading coefficients. Then $f_1(x) + \cdots + f_m(x)$ interlaces $c_1 f_1(x) + \cdots + c_m f_m(x)$ for all positive reals $c_1 \geq \cdots \geq c_m$. In particular, $f_1(x) + \cdots + f_{m-1}(x)$ interlaces $f_1(x) + 2f_2(x) + \cdots + m f_m(x)$.*

*Proof.* We only need to prove part (c) and for that, we proceed by induction on $m$. The case $m = 1$ being trivial, let us assume that the result holds for a positive integer $m - 1$, consider a sequence $(p_1(x), p_2(x), \ldots, p_m(x))$ and positive reals $c_1 \leq c_2 \leq \cdots \leq c_m$ as in the statement of the lemma and set $s_m(x) := p_1(x) + p_2(x) + \cdots + p_m(x)$. Since the sequence $(p_1(x), \ldots, p_{m-2}(x), p_{m-1}+p_m(x))$ is also interlacing [Fis08, Lemma 3.4], the induction hypothesis implies that $s_m(x)$ interlaces $c_1 p_1(x) + \cdots + c_{m-2}p_{m-2}(x) + c_{m-1}(p_{m-1}(x) + p_m(x))$. Since $s_m(x)$ also interlaces $(c_m - c_{m-1})p_m(x)$ (because each of its summands does so), it must interlace the sum of these two polynomials. This completes the induction.

For the second statement, let $s_{m-1}(x) := p_1(x) + p_2(x) + \cdots + p_{m-1}(x)$. Form the first statement we have that $s_{m-1}(x)$ interlaces $p_1(x) + 2p_2(x) + \cdots + (m - 1)p_{m-1}(x)$. Since $s_{m-1}(x)$ also interlaces $mp_m(x)$, it must interlace the sum of these two polynomials and the proof follows. $\square$

To prove real-rootedness of polynomials that satisfy a (linear) recurrence relation, one way to do so is to show that the defining recurrence relation preserves the real-rootedness of polynomials. This gives rise to the question of characterizing the linear operators on polynomials preserving real-rootedness. In [BB09], Borcea and Brandén provided the characterization of real-rootedness-preserving operators, based on the notion of *stable polynomials*, that extends the notion of real-rootedness to multivariate polynomials.

**Definition 1.58.** A polynomial $f(x_1, \ldots, x_m) \in \mathbb{C}[x_1, \ldots, x_m]$ is called *stable* if $\mathrm{Im}(x_1) > 0, \ldots, \mathrm{Im}(x_m) > 0$ implies $f(x_1, \ldots, x_m) \neq 0$.

So a univariate polynomial is real-rooted if and only if it is stable.

Now there is the following characterization of operators preserving real-rootedness.

**Theorem 1.59.** [BB09, Theorem 5] *Let $T : \mathbb{R}[x] \to \mathbb{R}[x]$ be a linear operator. Then $T$ preserves real-rootedness if and only if one of the following conditions is satisfied:*

1. *$T$ has rank at most two and is of the form*

$$T(p) = \alpha(p)f + \beta(p)g,$$

   *where $\alpha, \beta : \mathbb{R}[x] \to \mathbb{R}$ are linear functionals and $f, g$ are real-rooted, interlacing polynomials.*

2. *$G_T(x, y)$ is stable (with $G_T(x, y) = \sum_{k=0}^{n} \binom{n}{k} T(x^k)y^{n-k}$, called symbol of $T$).*

3. *$G_T(x, -y)$ is stable.*

This theorem also yields another proof for the real-rootedness of the Eulerian polynomial, since the linear operator given by the recursion for the Eulerian polynomials satisfies condition 2. of Theorem 1.59, see [Brä15, Example 7.3].

Another method for proving real-rootedness of families of polynomials is that of *compatible polynomials*.

**Definition 1.60.** The polynomials $f_1(x), \ldots, f_m(x)$ are called *k-compatible*, with $1 \le k \le m$ if

$$\sum_{j \in S} \lambda_j f_j(x)$$

is real-rooted whenever $S \subseteq [m]$, $|S| = k$ and $\lambda_j \ge 0$ for all $j \in S$.

Chudnovsky and Seymour proved the following theorem [CS07, Theorem 3.6], that relates the notions of compatible and interlacing polynomials.

**Theorem 1.61.** [CS07, Theorem 3.6] *Suppose that the leading coefficients of the polynomials $f_1(x), \ldots, f_m(x) \in \mathbb{R}[x]$ are positive. The following are equivalent:*

1. *$f_1(x), \ldots, f_m(x)$ are 2-compatible.*

2. *For all $1 \le i < j \le m$, $f_i(x)$ and $f_j(x)$ have a proper common interlacing polynomial.*

3. *$f_1(x), \ldots, f_m(x)$ have a proper common interlacing polynomial.*

4. *$f_1(x), \ldots, f_m(x)$ are m-compatible.*

Now we want to put the property of being alternatingly increasing into the picture. Every polynomial $f(x) \in \mathbb{R}[x]$ of degree at most $n$ can be written uniquely as $f(x) = a(x) + xb(x)$, where $a(x)$ and $b(x)$ are symmetric with centers of symmetry $\frac{n}{2}$ and $\frac{n-1}{2}$ respectively. This property is directly related to a polynomial being alternating increasing, since a polynomial is alternating increasing if and only if the the polynomials $a(x)$ and $b(x)$ are unimodal and have nonnegative coeffcents.

**Definition 1.62.** The polynomial $f(x)$ has a *nonnegative symmetric decomposition* with respect to $n$, if $a(x)$ and $b(x)$ have nonnegative coefficients. If in addition the polynomials $a(x)$ and $b(x)$ are real-rooted, then we say that $f(x)$ has a *real-rooted symmetric decomposition*. If additionaly also $x^n f(\frac{1}{x})$ interlaces $f(x)$, we say that $f(x)$ has a *real-rooted and interlacing symmetric decomposition*.

Brändén and Solus proved the following result on symmetric decompositions [BS19, Theorem 2.6]:

**Theorem 1.63.** *Let $f(x) \in \mathbb{R}[x]$ be a polynomial of degree at most $n$ and with a nonnegative symmetric decomposition $f(x) = a(x) + xb(x)$. Then the following are equivalent:*

1. *$b(x)$ interlaces $a(x)$.*

2. *$a(x)$ interlaces $f(x)$.*

3. *$b(x)$ interlaces $f(x)$.*

4. *$x^n f(\frac{1}{x})$ interlaces $f(x)$.*

The interlacing property in Theorem 1.63 implies, that all of the mentioned polynomials $f(x), a(x), b(x)$ are real-rooted, and hence log-concave and unimodal.

**Example 1.64.** Strengthening the results on the real-rootedness of the polynomials $A_{n,r}$ and $d_{n,r}$ mentionend in Example 1.52, Brändén and Solus prove that these polynomials are interlaced by their reciprocals [BS19, Corollary 3.2 and Theorem 3.8], and hence by Theorem 1.63 have a real-rooted and interlacing symmetric decomposition.

# 2 Combinatorics of subdivisions of simplicial complexes

An important question in the field of subdivisions of simplicial complexes is the behaviour of invariants, such as the $h$-vector, under a subdivision of the simplicial complex. Stanley and Kalai raised the question, if the $h$-vector of a Cohen–Macaulay complex increases under subdivisions. For a certain class of subdivisions, namely quasi-geometric subdivisons, Stanley answered it in the affirmative [Sta92]. A key tool for the study of subdivision that he introduced is the so-called local $h$-vector.

Starting with the barycentric subdivision in the original paper of Stanley [Sta92], the local $h$-vector has been studied for several different kinds of subdivisions, in order to find characterizations of different classes of subdivisions, e.g., see [Cha94], [JKMS19], or to give explicit combinatorial interpretations for certain subdivisions, as the $r$-th edgewise subdivision [Ath16a] and the interval subdivision [AS13], [AN18]. In [Ath16b], Athanasiadis gives a detailed introduction and overview on this topic.

Since the local $h$-vector is symmetric, it makes sense to define the $\gamma$-vector for the local $h$-vector, which is called local $\gamma$-vector. For several classical examples of subdivisions, such as the barycentric subdivision [AS12], the local $\gamma$-vector was shown to be nonnegative, and combinatorial interpretations of these vectors were given.

In the first part of this chapter, we start by introducing the basic definitions and some general results on subdivisions, the local $h$- and the local $\gamma$-vector.

In the second part, we define so-called *uniform triangulations*. This concept has been introduced recently in [Ath20b] to give a general framework and explain results for several examples of subdivisions, e.g., on the real-rootedness of $h$-polynomials, .

In the third part, we will consider explicit examples of subdivisions, namely the barycentric, edgewise and interval subdivision, and review known results for these subdivisions on the $h$-, the local $h$- and local $\gamma$-vector and their combinatorial interpretations.

## 2.1 Basic definitions on subdivisions of simplicial complexes

Following Stanley [Sta92], we define the subdivision of a simplicial complex.

**Definition 2.1.** A *topological subdivision* of a simplicial complex $\Delta$ is a pair $(\Gamma, \sigma)$, where $\Gamma$ is a simplicial complex and $\sigma : \Gamma \to \Delta$ is a map such that for any $F \in \Delta$:

(i) $\Gamma_F := \sigma^{-1}(2^F)$ is a subcomplex of $\Gamma$ which is homeomorphic to a ball of dimension $\dim(F)$. The complex $\Gamma_F$ is called the *restriction* of $\Gamma$ to $F$.

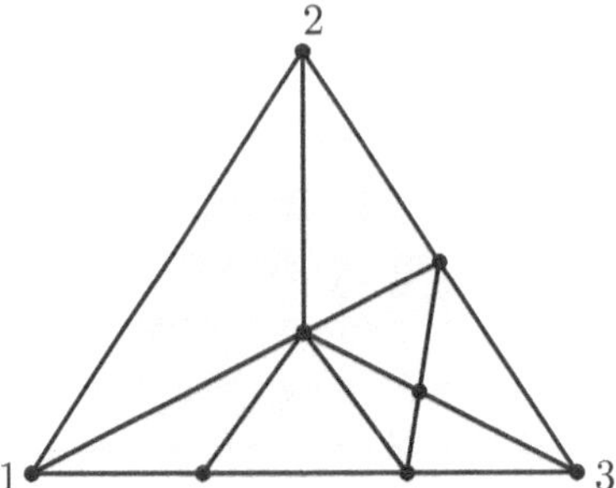

Figure 2.1: A subdivision of the 2-simplex

(ii) $\sigma^{-1}(F)$ consists of the interior faces of $\Gamma_F$.

The face $\sigma(G) \in \Delta$ is called the *carrier* of $G \in \Gamma$.

We notice that the subdivision map $\sigma$ is inclusion preserving, i.e., $\sigma(F) \subseteq \sigma(G)$ if $F \subseteq G$.

**Example 2.2.** The simplicial complex shown in Figure 2.1 is a subdivision of the 2-simplex.

There are several different classes of subdivisions, including the following:

**Definition 2.3.** Let $\Gamma$ be a subdivision $(\Gamma, \sigma)$ of a simplicial complex $\Delta$.

(i) $(\Gamma, \sigma)$ is called *quasi-geometric* if there does not exist $E \in \Gamma$ and $F \in \Delta$ with $\dim(F) < \dim(E)$ such that $\sigma(v) \subseteq F$ for any vertex $v$ of $E$.

(ii) $(\Gamma, \sigma)$ is called *vertex-induced*, if for all faces $E \in \Gamma$ and $F \in \Delta$ such that every vertex of $E$ is a vertex of $\Gamma_F$ we have $E \in \Gamma_F$.

(iii) $(\Gamma, \sigma)$ is called *geometric* if $\Gamma$ admits a geometric realization that geometrically sudivides a geometric realization of $\Delta$. Here we also say $\Gamma$ is a *triangulation* of $\Delta$.

(iv) $(\Gamma, \sigma)$ is called *regular* if the subdivision is induced by a weight function, i.e., it can be obtained via a projection of the lower hull of a polytope.

(v) $(\Gamma, \sigma)$ is called *flag subdivision* if the restriction $\Gamma_F$ is a flag complex for every face $F \in \Delta$.

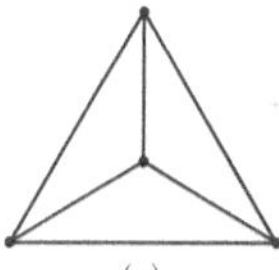 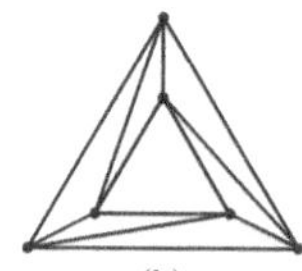 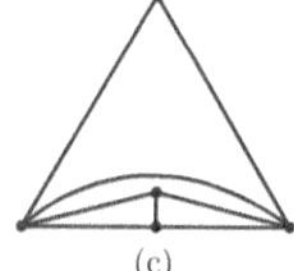 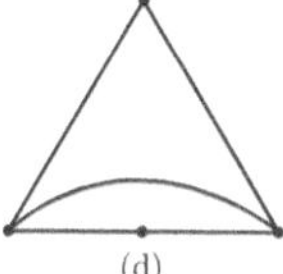

(a)      (b)      (c)      (d)

Figure 2.2: Examples of subdivisions of the 2-simplex

The first four classes of subdivisions are ordered in a way such that they are getting more restrictive, meaning they are related to each other as follows:

$$\{\text{topological subdivisions}\} \supsetneq \{\text{quasi-geometric subdivisions}\} \supsetneq$$
$$\{\text{vertex-induced subdivisions}\} \supsetneq \{\text{geometric subdivisions}\} \supsetneq \{\text{regular subdivisions}\}.$$

**Example 2.4.** In Figure 2.2 one can find several examples of subdivisions of the 2-simplex, also showing that some of the inclusions are strict. Subdivision (a) is a regular subdivision, (b) is vertex-induced but not geometric, (c) is quasi-geometric but not vertex-induced, while (d) is not even quasi-geometric. An example for a subdivision, that is vertex-induced but not geometric, can be found in [Cha94, Section 2].

In the following we define some classic and well studied subdivisions, the barycentric subdivision and the edgewise subdivision as well as the interval subdivision.

**Definition 2.5.** The *barycentric subdivision* $\mathrm{sd}(\Delta)$ of a simplicial complex $\Delta$ is the abstract simplicial complex whose $i$-dimensional faces are strictly increasing flags of the form $F_0 \subset \cdots \subset F_i$ where $F_j$, for $0 \leq j \leq i$, is a nonempty face of $\Delta$.

**Definition 2.6.** Suppose $\Delta$ is a simplicial complex on the vertex set $\Omega = \{e_1, \ldots, e_m\}$ of coordinate vectors of $\mathbb{R}^m$. For $a = (a_1, \ldots, a_m) \in \mathbb{Z}^m$ its *support*, denoted by $\mathrm{supp}(a)$, is the set of indices $i \in [m]$ for which $a_i \neq 0$ and write $i(a) = (a_1, a_1 + a_2, \ldots, a_1 + a_2 + \cdots + a_m)$. Now the *r-th edgewise subdivision* of $\Delta$ is the complex $\mathrm{esd}_r(\Delta)$ on vertex set $\Omega_r = \{(i_1, \ldots, i_m) \in \mathbb{N}^m : i_1 + \cdots + i_m = r\}$, and a set $G \subseteq \Omega_r$ is a face if the following two conditions are satisfied:

(i) $\bigcup_{u \in G} \mathrm{supp}(u) \in \Delta$,

(ii) $i(u) - i(v) \in \{0, 1\}^m$ or $i(v) - i(u) \in \{0, 1\}^m$ for all $u, v \in G$.

**Definition 2.7.** The *interval subdivision* $\mathrm{Int}(\Delta)$ of $\Delta$ is a simplicial complex on the vertex set $I(\Delta \setminus \emptyset)$ with

$$I(\Delta \setminus \emptyset) := \{[A, B] : \emptyset \neq A \subseteq B \in \Delta\},$$

as a poset ordered by inclusion, defined by $[A, B] \subseteq [C, D] \in I(\Delta \setminus \emptyset)$ if and only if $C \subseteq A \subseteq B \subseteq D$. $\mathrm{Int}(\Delta)$ is now the simplicial complex consisting of all chains in this poset.

30

These subdivisions will serve as examples for many observations on subdivisions, and their properties are reviewed in more detail in Section 2.4.

The main tool to study the behaviour of the $h$-vector under a subdivision is the local $h$-vector, introduced by Stanley [Sta92, Definition 2.1].

From now on and for the rest of the book, unless not stated otherwise explicitly, let $V$ be a set with $n$ elements.

**Definition 2.8.** Let $\Gamma$ be a homology subdivision of the simplex $2^V$. The polynomial $\ell_V(\Gamma, x) = \ell_0 + \ell_1 x + \cdots + \ell_n x^n$ is defined by

$$\ell_V(\Gamma, x) = \sum_{F \subseteq V} (-1)^{n-|F|} h(\Gamma_F, x)$$

and is called the *local $h$-polynomial* of $\Gamma$. The sequence $\ell_V(\Gamma) = (\ell_0, \ldots, \ell_n)$ is the *local $h$-vector* of $\Gamma$.

**Example 2.9.** As an example, we calculate the local $h$-polynomial of the subdivision $\Gamma$ in Figure 2.1. We have to calculate the $h$-polynomials of the restrictions $\Gamma_F$ for all $F \in \sigma_3$. We get $h(\Gamma_F, x) = 1$ whenever $F = \emptyset$ or $\dim(F) = 0$.

For the 1-dimensional faces we get $h(\Gamma_{\{1,2\}}, x) = 1$, $h(\Gamma_{\{2,3\}}, x) = 1 + x$ and $h(\Gamma_{\{1,3\}}, x) = 1 + 2x$.

For the 2-dimensional face $F = \{1, 2, 3\}$ we get $h(\Gamma_{\{1,2,3\}}, x) = 1 + 5x + 2x^2$.

So we get

$$\ell_V(\Gamma, x) = (1 + 5x + 2x^2) - (1 + (1 + x) + (1 + 2x)) + (1 + 1 + 1) - 1$$
$$= 2x^2 + 2x.$$

By the principle of inclusion-exclusion, Definition 2.8 also yields the following expression for the $h$-polynomial in terms of local $h$-polynomials:

$$h(\Gamma, x) = \sum_{W \subseteq V} \ell_W(\Gamma_W, x). \tag{2.1}$$

The local $h$-vector has the following nice properties, summarizing the results [Sta92, Theorem 3.2, Theorem 3.3, Theorem 5.2 and Corollary 4.7].

**Theorem 2.10.** [Sta92]

(i) *Let $\Delta$ be a simplicial complex and let $\Gamma$ be a subdivision of $\Delta$. Then:*

$$h(\Gamma, x) = \sum_{F \in \Delta} \ell_F(\Gamma_F, x) h(\mathrm{lk}_\Delta(F), x). \tag{2.2}$$

(ii) *Let $V \neq \emptyset$ and let $\Gamma$ be a subdivision of $2^V$. Then:*

a) *The local $h$-vector is symmetric, i.e., $\ell_i(\Gamma) = \ell_{n-i}(\Gamma)$ for $0 \le i \le n$. Furthermore, $\ell_0(\Gamma) = 0$ and $\ell_1(\Gamma) \ge 0$.*

*b) If $\Gamma$ is quasi-geometric, then $\ell_i(\Gamma) \geq 0$ for $0 \leq i \leq n$.*

*c) If $\Gamma$ is regular, then $\ell_V(\Gamma)$ is unimodal.*

For proving the nonnegativity of the local $h$-vector, Stanley finds certain graded modules related to the subdivision $\Gamma$ in the following way: First, we call a homogeneous system of parameters (h.s.o.p for short, see also Section 4.1.2 for the definition of an l.s.o.p.) $\theta_1, \ldots, \theta_n$ *special*, if each $\theta_i$ is a linear combination of variables $x_j$ corresponding to vertices, which are not in the closure of the face $V \smallsetminus \{x_i\}$. Now Stanley shows, that for $\mathbb{F}[\Gamma]$ there exists a special h.s.o.p. if and only if $\Gamma$ is a quasi-geometric subdivision [Sta92, Corollary 4.4]. Call the image of the ideal of $\mathbb{F}[\Gamma]$, which is generated by the interior faces $\Gamma$, in $\mathbb{F}[\Gamma]/(\theta_1, \ldots, \theta_n)$ with a special h.s.o.p. $(\theta_1, \ldots, \theta_n)$ the *local face module* of $\Gamma$. This module is a graded ideal, and Stanley shows that the dimensions of its graded parts are equal to the entries of the local $h$-vector [Sta92, Theorem 4.6]. Hence they are greater than or equal to 0.

With the different properties stated in Theorem 2.10($ii$) in mind, one can ask for characterizations of local $h$-vectors of different kinds of subdivisions.

The properties in Theorem 2.10($ii$)a) already characterize the local $h$-vectors of topological subdivisions, shown in [Cha94, Theorem 4.1]. Furthermore it was shown, that for a characterization of regular subdivisions one has to add unimodality. Stanley and Chan (see [Sta92, Conjecture 5.4], [Cha94, Section 6]) also conjectured the unimodality of the local $h$-vector for every quasi-geometric subdivision. But in [Ath12, Example 3.4] a counterexample was given, and it was shown in [JKMS19, Theorem 1.1], that the properties in ($ii$)a) and ($ii$)b) of Theorem 2.10 characterize the local $h$-vectors of quasi-geometric subdivisions. But in view of no known example of a vertex-induced subdivision with a local $h$-vector, that is not unimodal, Athanasiadis [Ath12, Question 3.5] raised the question, if all vertex-induced subdivisions have a unimodal local $h$-vector.

When studying the different terms on the right-hand side of (2.2), one observes that the term for $F = \emptyset$ is equal to $h(\Delta, x)$. Furthermore, from part ($ii$)b) of Theorem 2.10 we see that for a Cohen–Macaulay simplicial complex $\Delta$ and a quasi-geometric subdivision $\Gamma$ all polynomials appearing as factors in the sum in (2.2) have nonnegative coefficients, and so has the $h$-polynomial as the sum. By this one obtains the following result, giving a partial answer to the question about the behaviour of the $h$-vector under subdivision.

**Theorem 2.11.** [Sta92, Theorem 4.10] *Let $\Delta$ be a Cohen–Macaulay simplicial complex and $\Gamma$ be a quasi-geometric subdivision of $\Delta$. Then*

$$h(\Gamma) \geq h(\Delta).$$

Since by Theorem 2.10 the local $h$-vector is symmetric, we can consider the $\gamma$-vector of the local $h$-polynomial, as for general symmetric polynomials in Chapter 1.3.

**Definition 2.12.** Let $\Gamma$ be a subdivision of $2^V$. One can express $\ell_V(\Gamma, x)$ as

$$\ell_V(\Gamma, x) = \sum_{j=0}^{\lfloor \frac{n}{2} \rfloor} \xi_j(\Gamma) x^j (1 + x)^{n-2j},$$

where $\xi_j(\Gamma) \in \mathbb{Z}$ is uniquely determined. The sequence $\xi_V(\Gamma) = (\xi_0(\Gamma), \ldots, \xi_{\lfloor \frac{n}{2} \rfloor}(\Gamma))$ is called the *local $\gamma$-vector of $\Gamma$*.

For Eulerian complexes, one gets a similar description of the $\gamma$-vector in terms of local $\gamma$-vectors, as we have seen it for the $h$-vector in terms of the local $h$-vectors in (2.2).

**Theorem 2.13.** [Ath12, Proposition 5.3] *Let $\Delta$ be an Eulerian simplicial complex, and $\Gamma$ be a homology subdivision of $\Delta$. Then*

$$\gamma(\Gamma, x) = \sum_{F \in \Delta} \xi_F(\Gamma_F, x) \gamma(\mathrm{lk}_\Delta(F), x). \tag{2.3}$$

## 2.2 Gal's conjecture and beyond

An important and classical problem in combinatorics is to characterize the face vectors of simplicial complexes. In Chapter 1.1.4 we saw, that the characterization of $f$-vectors for simplicial complexes is given by the Kruskal-Katona inequalities, and for balanced simplicial complexes it is provided by the Frankl-Füredi-Kalai inequalities. Nevertheless, one can consider further classes of complexes and try to characterize their $f$- and $h$-vectors.

Th aim of this section is to review some of the main conjectures on characterizations for the class of flag simplicial spheres, which also gives a motivation to study the $h$-vectors and $\gamma$-vectors and their combinatorial interpretations of subdivisions of simplicial complexes. Following [Sie17, Remark 3.2.5] and [Pet15, Section 10.8] we state some conjectures and sketch the connection between these conjectures, as well as state some of the known cases.

Motivated by a conjecture of Hopf (see for example [Che66]) on the Euler-characteristic of manifolds, Charney and Davis [CD95, Conjecture D] conjectured the following nonnegativity statement for the $h$-polynomial of flag spheres.

**Conjecture 2.14.** [CD95] *If $\Delta$ is a flag simplicial sphere of dimension $n - 1 = 2m - 1$, then*

$$(-1)^m h(\Delta, -1) \geq 0.$$

We remark, that the Charney-Davis conjecture would follow from the real-rootedness of the $h$-polynomial of the flag sphere.

Since $\Delta$ is a sphere, its $h$-polynomial is symmetric, and as before we can write it as

$$h(\Delta, x) = \sum_{k=0}^{\lfloor \frac{n}{2} \rfloor} \gamma_k(\Delta) x^k (1 + x)^{n-2k}$$

and study its $\gamma$-vector $\gamma(\Delta) = (\gamma_0(\Delta), \ldots, \gamma_{\lfloor \frac{n}{2} \rfloor}(\Delta))$. So by the definition of the $\gamma$-vector we get

$$h(\Delta, -1) = \begin{cases} (-1)^m \gamma_m(\Delta), & \text{if } n = 2m \\ 0, & \text{otherwise.} \end{cases}$$

So the Charney-Davis conjecture follows if $\gamma_{\frac{n}{2}}(\Delta) \geq 0$ whenever $n$ is even.

Generalizing this observation and strengthening the Charney-Davis conjecture, Gal [Gal05, Conjecture 2.1.7] raised the conjecture, that every entry of the $\gamma$-vector is nonnegative for every flag sphere.

**Conjecture 2.15.** [Gal05] *If $\Delta$ is a flag simplicial sphere, then $\gamma(\Delta)$ is nonnegative.*

This statement was proven for 3-dimensional spheres [DO01, Theorem 11.2.1] and also 4-dimensional spheres [Gal05, Corollary 2.2.3]. It is also known for several other classes of spheres, such as finite Coxeter complexes [Ste08, Therorem 1.2], chordal nestohedra [PRW08, Theorem 1.1], which include for example associahedra and permutohedra, and several subdivisions of spheres, see Section 2.4.

Gal also considered a further strengthening of this conjecture, that the $h$-polynomial of such a sphere is real-rooted, which would imply its $\gamma$-nonnegativity. This statement turns out to be false in general. It is true for flag spheres of dimension up to four [Gal05, Theorem 3.1.3], but one can construct flag simplicial spheres of dimension five with an $h$-polynomial, that is not real-rooted, but its $\gamma$-vector is still nonnegative [Gal05, Section 3.3].

Postnikov, Reiner and Williams [PRW08, Conjecture 14.2] extend Gal's conjecture, by conjecturing the monotonicity of the $\gamma$-vector of geometric subdivisions $\Gamma$ of homology spheres. This can be seen as a conjectural analogue of Theorem 2.11.

Athanasiadis [Ath12, Theorem 1.3] gave an answer to this conjecture in the affirmative in the case of vertex-induced subdivisions of 3- and 4-dimensional flag spheres, which led to the following stronger version of the conjecture, raised by Athanasiadis.

**Conjecture 2.16.** [Ath12, Conjecture 1.4] *Let $\Delta$ be a flag simplicial sphere and $\Gamma$ be a vertex-induced homology subdivision of $\Delta$. Then $\gamma(\Gamma) \geq \gamma(\Delta)$.*

Since the local $h$-vector is symmetric and one can define the local $\gamma$-vector, it is quite natural to ask analogous questions for this invariant. Here the main conjecture was raised by Athanasiadis, who also gave an affirmative answer to it in the case of subdivisions in dimension less than or equal to 3 [Ath12, Proposition 5.7].

**Conjecture 2.17.** [Ath12, Conjecture 5.4] *For every flag vertex-induced subdivision $\Gamma$ of the simplex $2^V$ the local $\gamma$-vector of $\Gamma$ is nonnegative, i.e., $\xi_V(\Gamma) \geq 0$.*

As we will see in Section 2.4, this conjecture is known to be true for several special classes of subdivisions, including the barycentric subdivision.

Conjecture 2.17 is directly related to Gal's conjecture. Similar as in the reasoning for the monotonicity of the $h$-vector, under the assumption that $\gamma(\mathrm{lk}_\Delta(F))$ is nonnegative, the nonnegativity of the local $\gamma$-vector together with (2.3) implies the monotonicity of the $\gamma$-vector.

But actually there is another way, that is not using the assumption on the $\gamma$-vector of the links: The fact, that each flag homology sphere $\Delta$ can be seen as a vertex-induced subdivision of the boundary complex $C_{n-1}$ of the cross-polytope [Ath12, Theorem 1.5], yields the statement [Ath12, Corollary 5.5] that the $\gamma$-vector of $\Delta$ can be described by

$$\gamma(\Delta, x) = \sum_{F \in C_{n-1}} \xi_F(\Delta_F, x), \tag{2.4}$$

where $\Delta_F$ is some flag vertex-induced subdivision of the simplex $2^F$ for each $F \in C_{n-1}$. Hence the nonnegativity of the local $\gamma$-vector of flag vertex-induced subdivisions of the simplex together with (2.4) would imply the nonnegativity of the $\gamma$-vector of flag simplicial spheres, showing that Conjecture 2.17 implies Gal's conjecture. Furthermore, using the arguments from this discussion, Conjecture 2.17 also implies Conjecture 2.16 [Ath12, Corollary 5.6]. Summarizing, we get

$$\text{Conj 2.17} \Rightarrow \text{Conj. 2.16} \Rightarrow \text{Conj. 2.15} \Rightarrow \text{Conj. 2.14.}$$

Nevo and Petersen [NP11, Theorem 1.2 and 6.1] show for several families of flag homology spheres, that their $\gamma$-vectors satisfy the Kruskal-Katona inequalities and the even stronger Frankl-Füredi-Kalai inequalities. In regard of these results, they conjecture the following combinatorial interpretation of the $\gamma$-vector, which clearly implies its nonnegativity and hence would also imply Gal's Conjecture 2.15.

**Conjecture 2.18.** [NP11, Problem 6.4] *The $\gamma$-vector of a flag simplicial sphere is the $f$-vector of a balanced simplicial complex, i.e., $\gamma(\Delta)$ satisfies the Frank-Füredi-Kalai inequalities whenever $\Delta$ is a flag sphere.*

## 2.3 Uniform triangulations

Many of the well known subdivisions share similar properties for many invariants. As we will see in the next section for several examples of subdivisions, for instance one obtains the $h$-vector of the subdivision by a linear transformation of the $h$-vector of the original complex, and one can interpret the coefficients arising in this transformation by counting combinatorial objects. Hence these numbers are nonnegative, as it was shown for the barycentric subdivision [BW08] and the interval subdivision [AN20]. Another property of interest is the real-rootedness of the $h$-polynomial, which then gives the log-concavity und unimodality of the $h$-vector. Again, it is shown for barycentric subdivisions [BW08] and interval subdivisions [AN20] to have real-rooted $h$-polynomials. Furthermore, these subdivisions have in common that restricting the subdivision to an $i$-dimensional face of the original complex, yields again the same kind of subdivision of the $i$-dimensional simplex.

Motivated by these facts, Athanasiadis [Ath20b] introduced a general framework for these special classes of subdivisions, the so-called uniform triangulations.

An $f$-*triangle* of size $d$ is a triangular array $\mathcal{F} = (f_{\mathcal{F}}(i,j))_{0 \leq i \leq j \leq d}$ of nonnegative integers. We call

$$f_{\mathcal{F}}(\sigma_n, x) := \sum_{i=0}^{n} f_{\mathcal{F}}(i,n)x^i$$

the $n$-*th $f$-polynomial* associated to $\mathcal{F}$.

**Definition 2.19.** Let $\mathcal{F}$ be an $f$-triangle of size $d$ and let $\Delta$ be a simplicial complex of dimension less than $n$. A triangulation $\Gamma$ of $\Delta$ is called $\mathcal{F}$-*uniform* if we have $f(\Gamma_F, x) = f_{\mathcal{F}}(\sigma_n, x)$ for every $(n-1)$-dimensional face $F \in \Delta$ and all $n \leq d$.

Another, equivalent property is that for all $0 \leq i \leq j \leq d$ the restriction of $\Gamma$ to any $(j-1)$-dimensional face of $\Delta$ has exactly $f_{\mathcal{F}}(i,j)$ many $(i-1)$-dimensional faces. If $\Gamma$ is an $\mathcal{F}$-uniform triangulation of $\Delta$, then so is the restriction of $\Gamma$ to any subcomplex of $\Delta$.
We will call an $\mathcal{F}$-uniform triangulation also just *uniform triangulation* for short.

Important examples of subdivisions, that fall into the class of uniform triangulations, are the barycentric subdivision and the $r$-th edgewise subdivision, which will be discussed in greater detail in the upcoming Chapters 2.4.1 and 2.4.2 respectively.

Generalizing known statements on the $h$-vector transformation, as for the barycentric [BW08] or interval subdivision [AN20], there is the following result on the $h$-vector transformation of uniform triangulations.

**Theorem 2.20.** [Ath20b, Theorem 1.1] *Let $\mathcal{F}$ be a $f$-triangle of size $d$. There exist nonnegative integers $p_{\mathcal{F}}(n,k,j)$ for $n \in [d]$ and $k, j \in [n]$ such that for all $n \leq d$*

$$h_j(\Gamma) = \sum_{k=0}^{n} p_{\mathcal{F}}(n,k,j)h_k(\Delta)$$

*for every $(n-1)$-dimensional simplicial complex $\Delta$, every $\mathcal{F}$-uniform triangulation $\Gamma$ of $\Delta$ and all $j \in n$.*

The nonnegativity of the numbers $p_{\mathcal{F}}(n,k,j)$ follows by expressing them as a nonnegative linear combination of coefficients of certain local $h$-polynomials arising from the usual transformation from the $f$- to the $h$-polynomial and the structure of $f$-vectors of uniform triangulations.
Not only the nonnegativity of the numbers $p_{\mathcal{F}}(n,k,j)$ was proven, but also several other properties [Ath20b, Proposition 4.6], such as the symmetry

$$p_{\mathcal{F}}(n,k,j) = p_{\mathcal{F}}(n,n-k,n-j) \tag{2.5}$$

and the recursion

$$p_{\mathcal{F}}(n,k,j) = p_{\mathcal{F}}(n,k-1,j) + p_{\mathcal{F}}(n-1,k-1,j-1) - p_{\mathcal{F}}(n-1,k-1,j). \quad (2.6)$$

This recursion was proven by using the linear operators describing the transformation of the $f$- and $h$-vector under a uniform transformation [Ath20b, Corollary 5.6]. In Section 3.5 we provide another direct proof for this recurrence in the case of the antiprism triangulation.

Athanasiadis also gives a criterion to prove the real-rootedness of the $h$-polynomial of the uniform triangulation of a simplicial complex.

As in [Ath20b], for a given $\mathcal{F}$-triangle, we denote the $h$-polynomial of any $\mathcal{F}$-uniform triangulation of the simplex $\sigma_n$ and its boundary by $h_{\mathcal{F}}(\sigma_n, x)$ and $h_{\mathcal{F}}(\partial(\sigma_n), x)$, respectively. Similarly, by $\ell_{\mathcal{F}}(\sigma_n, x)$ we denote the local $h$-polynomial of any $\mathcal{F}$-uniform triangulation of the simplex $\sigma_n$.

**Theorem 2.21.** [Ath20b, Theorem 1.2] *Let $\mathcal{F}$ be an $f$-triangle of size $d \in \mathbb{N}$ and $\Gamma_n$ be an $\mathcal{F}$-uniform triangulation of the $(n-1)$-dimensional simplex. Assume the following:*

*(i) $h(\Gamma_n, x)$ is a real-rooted polynomial for all $n < d$,*

*(ii) $h(\Gamma_n, x) - h(\partial(\Gamma_n), x)$ is either identically zero or a real-rooted polynomial of degree $n-1$ with nonnegative coefficients which is interlaced by $h(\Gamma_{n-1}, x)$ for all $n \leq d$.*

*Then for every $(d-1)$-dimensional simplicial complex $\Delta$ with nonnegative $h$-vector the polynomial $h(\Delta', x)$ is real-rooted for every $\mathcal{F}$-uniform triangulation $\Delta'$ of $\Delta$.*

The main tool to prove this theorem is the theory of interlacing polynomials. By using the recurrence (2.6) for the numbers $p_{\mathcal{F}}(n,k,j)$, certain sequences of polynomials are defined and then it is shown by induction, that these are interlacing sequences and one can deduce real-rootedness.

As we will see in the next section, this result can be used to obtain already known results on real-rootedness of $h$-polynomials for some classes of subdivisions.

## 2.4 Examples of subdivisions

In this section we review combinatorial interpretations of invariants as the $h$-, local $h$- and local $\gamma$-vector of some examples of subdivisions of simplicial complexes, as well as known statements on the real-rootedness of the associated polynomials.

### 2.4.1 The barycentric subdivision

One of the most studied examples for subdivisions is the so-called barycentric subdivision, see Definition 2.5. In the following we give several other descriptions of the barycentric subdivision.

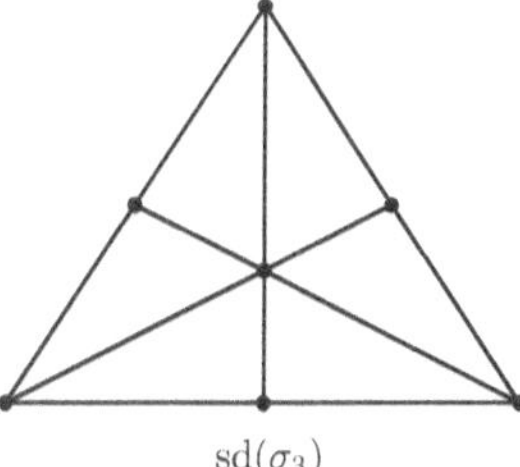


Figure 2.3: The barycentric subdivision of the 2-simplex

Consider the (simple, undirected) graph $\mathcal{G}(\Delta)$ on the node set of nonempty faces of $\Delta$ for which two nodes are adjacent if one is contained in the other. The barycentric subdivision sd($\Delta$) is defined as the *clique complex* of $\mathcal{G}(\Delta)$, meaning the abstract simplicial complex whose vertices are the nodes of $\mathcal{G}(\Delta)$ and whose faces are the sets consisting of pairwise adjacent nodes.

Geometrically, sd($\Delta$) can be described as a triangulation of $\Delta$ as follows. Assume that all faces of $\Delta$ of dimension at most $j$ have been triangulated, for some $j \in \mathbb{N}$. Then, triangulate each $(j+1)$-dimensional face of $\Delta$ by inserting one point in the interior of that face and coning over its boundary, which is already triangulated. By repeating this process, starting at $j = 0$ and moving to higher dimensional faces, we get a triangulation of $\Delta$ which is combinatorially isomorphic to sd($\Delta$).

Alternatively, sd($\Delta$) can be constructed by applying successively the operation of stellar subdivision to each face of $\Delta$ of positive dimension, starting from the facets and moving to lower dimensional faces in any order which respects reverse inclusion.

There is also another nice geometric definition of the barycentric subdivision of the simplex by intersecting the geometric standard simplex with cones coming from a certain hyperplane arrangement. The explicit description of this construction can be found for example in [AW18, Section 2.1].

**Example 2.22.** Figure 2.3 shows the barycentric subdivision of the simplex $\sigma_3$. The facets of the barycentric subdivision of the 2-simplex are given by

$$\{\{1\} \subset \{1,2\} \subset \{1,2,3\}, \{1\} \subset \{1,3\} \subset \{1,2,3\}, \{2\} \subset \{1,2\} \subset \{1,2,3\}$$
$$\{2\} \subset \{2,3\} \subset \{1,2,3\}, \{3\} \subset \{1,3\} \subset \{1,2,3\}, \{3\} \subset \{2,3\} \subset \{1,2,3\}\}.$$

One observes that the facets of the complex sd($\sigma_n$) can be encoded by all the permutations $\sigma \in \mathfrak{S}_n$, hence it has $n!$ facets.

We show, that the barycentric subdivision is an example of a uniform triangulation. For this we need to count all $i$-faces $F_0 \subset \cdots \subset F_i$ in sd($\Delta$), with a fixed face $F_i$ in $\Delta \smallsetminus \{\emptyset\}$. We can identify such a flag with $(F_0, F_1 \smallsetminus F_0, \ldots, F_i \smallsetminus F_{i-1})$, which gives a bijection between the $i$-faces of $\mathrm{sd}(\Delta)$ with $F_i$ the maximal element in the flag, and ordered set partitions of $F_i$ into $i+1$ many nonempty blocks. There are $(i+1)! S(\dim(F_i)+1, i+1)$ many of these partitions. So we get that the barycentric subdivision is a uniform triangulation with $f$-triangle $\mathcal{F} = (i! S(j, i))_{i \leq j}$.

Furthermore, it is clear, that the restriction of $\mathrm{sd}(\Delta)$ to a $j$-dimensional face $F \in \Delta$ is again the barycentric subdivision of the $j$-dimensional simplex $2^F$.

There is the following relation between the $f$-vectors of the barycentric subdivision $\mathrm{sd}(\Delta)$ and the simplicial complex $\Delta$, which is obtained by the same reasoning as above, see for example [BW08, Lemma 1].

**Theorem 2.23.** *Let $\Delta$ be an $(n-1)$-dimensional simplicial complex. Then*

$$f_j(\mathrm{sd}(\Delta)) = \sum_{i=0}^{n} f_{i-1}(\Delta)(j+1)! S(i, j+1)$$

*for all $-1 \leq j \leq n-1$.*

Before giving the $h$-vector transformation for an arbitrary simplicial complex, we consider the special case of the simplex $\sigma_n$. It holds that the barycentric subdivision of the simplex is the cone of the barycentric subdivision of the boundary of the simplex, i.e., $\mathrm{sd}(\sigma_n) = u * \mathrm{sd}(\partial(\sigma_n))$. Since coning just adds a zero at the end to the $h$-vector, we also get $h(\mathrm{sd}(\sigma_n)) = (h(\mathrm{sd}(\partial(\sigma_n))), 0)$, in particular the $h$-polynomials are the same. Now compute the $h$-vector of $\mathrm{sd}(\partial(\sigma_n))$ in the following way: First, one can associate each face of the boundary to a flag of the form $F_0 \subset F_1 \subset \cdots \subset F_i \subset [n]$, and by the reasoning as seen before, we get

$$f_{j-1}(\mathrm{sd}(\partial(\sigma_n))) = (j+1)! S(n, j+1).$$

Now using (1.3) and the relation between the Stirling numbers and the Eulerian numbers yields $h_j(\mathrm{sd}(\partial(\sigma_n))) = A(n, j)$ and hence

$$h(\mathrm{sd}(\sigma_n), x) = h(\mathrm{sd}(\partial(\sigma_n)), x) = A_n(x). \tag{2.7}$$

Brenti and Welker [BW08, Theorem 1] generalized (2.7) by giving the relation between the $h$-vector of an arbitrary simplicial complex $\Delta$ and its barycentric subdivision $\mathrm{sd}(\Delta)$ and providing a combinatorial interpretation of the numbers $p_{\mathcal{F}}(n, i, k)$ in the general framework of Theorem 2.20.

**Theorem 2.24.** [BW08] *Let $\Delta$ be an $(n-1)$-dimensional simplicial complex. Then*

$$h_j(\mathrm{sd}(\Delta)) = \sum_{i=0}^{n} A(n+1, j, i+1) h_i(\Delta)$$

*for all $0 \leq j \leq n$, where $A(n, j, i)$ is the number of permutations $\sigma \in \mathfrak{S}_n$ such that $\mathrm{des}(\sigma) = j$ and $\sigma(1) = i$.*

We remark that all the results in [BW08] were proven in greater generality, namely not just for simplicial complexes but for the more general class of *Boolean cell complexes*, which are CW-complexes for which the lower interval $[\emptyset, F]$ is a boolean lattice for each cell $F$ (see also [BW08, Chapter 1]).

From Theorem 2.24 it follows that

$$h_j(\mathrm{sd}(\Delta)) = A(n+1, j, j+1)h_j(\Delta) + \sum_{i \in [n] \setminus \{j\}} A(n+1, j, i+1)h_i(\Delta).$$

Under the assumption that the $h$-vector of $\Delta$ is nonnegative, one can conclude that $h_j(\mathrm{sd}(\Delta) \geq A(n+1, j, j+1)h_j(\Delta)$. Since $A(n+1, j, j+1) \geq 1$, one obtains the following corollary on the monotonicity of the $h$-vector under the barycentric subdivision.

**Corollary 2.25.** [BW08, Corollary 1] *Let $\Delta$ be an $(n-1)$-dimensional simplicial complex such that $h_j(\Delta) \geq 0$ for all $0 \leq j \leq n$. Then*

$$h_j(\mathrm{sd}(\Delta)) \geq h_j(\Delta)$$

*for $0 \leq j \leq n$.*

One can extend these results by considering the partial barycentric subdivision, which was done in [AW18]. Also in this case a combinatorial interpretation [AW18, Theorem 3.4] of the $h$-vector transformation was proven, generalizing Theorem 2.24.

By proving that the $h$-polynomial is real-rooted [BW08, Theorem 2], Brenti and Welker also obtained the log-concavity and hence unimodality of the $h$-polynomial of the barycentric subdivision. To show this, they related the appearing permutations in Theorem 2.24 to signed permutations and then used results on operators preserving real-rootedness of polynomials.

**Theorem 2.26.** [BW08] *Let $\Delta$ be an $(n-1)$-dimensional simplicial complex such that $h_i(\Delta) \geq 0$ for all $0 \leq i \leq n$. Then*

$$h(\mathrm{sd}(\Delta), x) = \sum_{i=0}^{n} h_i(\mathrm{sd}(\Delta))x^i$$

*has only real and simple roots.*
*In particular, $h(\mathrm{sd}(\Delta)) = (h_0(\mathrm{sd}(\Delta)), \ldots, h_n(\mathrm{sd}(\Delta)))$ is a log-concave and unimodal sequence.*

As it was mentioned in [Ath20b], this result can be recovered using Theorem 2.21 for the general setting of uniform triangulations in the following way. The $h$-polynomials of $\mathrm{sd}(\sigma_n)$ and $\partial(\mathrm{sd}(\sigma_n))$ are equal, see (2.7). Hence the second assumption in Theorem 2.21 is fulfilled since

$$h(\mathrm{sd}(\sigma_n), x) - h(\partial(\mathrm{sd}(\sigma_n)), x) = 0.$$

Now using Theorem 2.21, an induction proves the result.

Also for the local $h$-polynomial of the barycentric subdivision there is a combinatorial interpretation, given by Stanley [Sta92, Proposition 2.4]. By giving a bijection between permutations with a given set of fixed points and derangements and then using Equation (2.1), he proves the following result.

**Theorem 2.27.** [Sta92, Proposition 2.4] *The local $h$-polynomial of the barycentric subdivision of the $(n-1)$-dimensional simplex $2^V$ is given by*

$$\ell_V(\mathrm{sd}(2^V), x) = \sum_{w \in D_n} x^{\mathrm{exc}(w)} = d_n(x).$$

The nonnegativity of the local $\gamma$-vector of the barycentric subdivision and combinatorial interpretations were given in several different contexts (see [Ath16b, Chapter 4] and references therein). It was for example shown by Foata and Strehl [FS74], using a method now called 'valley-hopping'. With the help of a modified version of this method, see [AS12, Section 4], Athanasiadis and Savvidou could show the following interpretations of the local $\gamma$-vector.

**Theorem 2.28.** [AS12, Theorem 1.4] *The local $h$-polynomial of the barycentric subdivision of the $(n-1)$-dimensional simplex can be expressed as*

$$\ell_V(\mathrm{sd}(2^V), x) = \sum_{i=0}^{\lfloor \frac{n}{2} \rfloor} \xi_{n,i} x^i (1+x)^{n-2i},$$

*where $\xi_{n,i}$ denotes*

*(i) the number of permutations $w \in \mathfrak{S}_n$ with $i$ descending runs and no descending run of size one,*

*(ii) the number of derangements $w \in D_n$ with $i$ excedances and no double excedance,*

*(iii) the number of permutations $w \in \mathfrak{S}_n$ with $i$ descents and no double descent such that every left to right maximum of $w$ is a descent.*

We see, that by this result Conjecture 2.17 is confirmed for the barycentric subdivision.
Furthermore, the real-rootedness of the polynomial $\ell_V(sd(2^V), x)$ is clear, since we already discussed the real-rootedness of the derangement polynomial in Example 1.52.

Nevo, Petersen and Tenner show, that summing certain FFK-vectors, i.e., vectors satisfying the Frankl-Füredi-Kalai inequalities, yields an FFK-vector. They construct a balanced complex with an FFK-vector as $f$-vector by joining other certain balanced complexes and using the before mentioned result. Now refining the numbers $A(n, j, k)$ from Theorem 2.24, they show that the $\gamma$-vector of the barycentric subdivision equals the $f$-vector of the constructed balanced complex.

**Theorem 2.29.** [NPT11, Theorem 1.1] *If $\Delta$ is a simplicial complex with nonnegative and symmetric h-vector, then the $\gamma$-vector of the barycentric subdivision* $\mathrm{sd}(\Delta)$ *of $\Delta$ is the f-vector of a balanced simplicial complex.*

Hence the barycentric subdivision is one of the known classes, for which Conjecture 2.18 holds.

### 2.4.2 The $r$-th edgewise subdivision

Another classic subdivision is the $r$-th edgewise subdivision, given in Definition 2.6. There is also a nice geometrically way to describe this subdivision by cutting the geometric simplex with certain hyperplanes, it can be found in [Ath18, Section 3.3.1] for example. This subdivision has the property, that it subdivides each $j$-dimensional face of $\Delta$ into $r^j$ many $j$-dimensional faces in the subdivision $\mathrm{esd}_r(\Delta)$.

**Example 2.30.** The 4-th edgewise subdivision of the 2-simplex with some vertices labeled is depicted in Figure 2.4. The vertex set of this subdivision is $\Omega_4 = \{(i_1, i_2, i_3) \in \mathbb{N}^3 : i_1 + i_2 + i_3 = 4\}$. For example, this complex contains the edge $\{(0,2,2),(0,1,3)\} \in \mathrm{esd}_4(\sigma_3)$, since $i(0,2,2) = (0,2,4)$ and $i(0,1,3) = (0,1,4)$ and so $i(0,2,2) - i(0,1,3) = (0,1,0) \in \{0,1\}^3$.

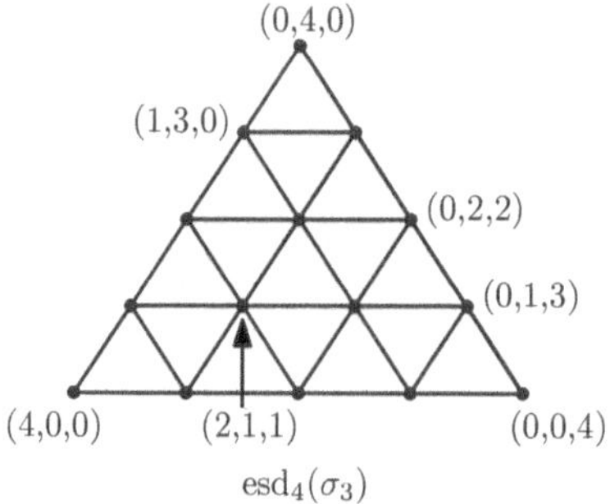


Figure 2.4: The 4-th edgewise subdivision of the 2-simplex

For the $h$-vector of the $r$-th edgewise subdivision one obtains the following result, by connecting the Stanley-Reisner ring of this subdivision to an algebraic object called Veronese algebra, see [BR05] and [BW09].

**Theorem 2.31.** [Ath18, Example 4.4] *Let $\Delta$ be a simplicial complex of dimension $n-1$. Then*

$$h(\mathrm{esd}_r(\Delta), x) = E_r((1 + x + x^2 + \cdots + x^{r-1})^n h(\Delta, x)),$$

*where $E_r : \mathbb{R}[x] \to \mathbb{R}[x]$ is the linear operator defined by setting $E_r(x^k) = x^{\frac{k}{r}}$ if $k$ is divisible by $r$, and $E_r(x^k) = 0$ otherwise.*

The real-rootedness of the $h$-polynomial of the edgewise subdivision follows from a result of Jochemko [Joc18, Theorem 1.1] on the Hilbert series of the Veronese algebra using interlacing properties. It can also be recovered using the methods for uniform triangulations, see [Ath20b, Example 7.2].

**Theorem 2.32.** [Joc18] *Let $\Delta$ be an $(n-1)$-dimensional simplicial complex such that $h_i(\Delta) \geq 0$ for all $0 \leq i \leq n$. If $r \geq n$, then $h(\mathrm{esd}_r(\Delta), x)$ has only real roots.*

When $r < \dim(\Delta)$, the $h$-polynomial of $\mathrm{esd}_r(\Delta)$ is not real-rooted in general, as it was mentioned in [BW09, Section 4] that the $h$-polynomial of $\mathrm{esd}_2(\partial(\sigma_5))$ is not real-rooted.

For a combinatorial interpretation of the local $h$-polynomial of the edgewise subdivision of the simplex, we need the so-called *Smirnov-words*: Denote by $S(n, r)$ the set of sequences $w = (w_0, \ldots, w_n) \in \{0, 1, \ldots, r-1\}^{n+1}$ having no two consecutive equal entries and satisfying $w_0 = w_n = 0$. Similar as defined for permutations in Chapter 1.2, we call an index $i \in \{0, \ldots, n\}$ an *ascent* if $w_i < w_{i+1}$ and denote the number of ascents of $w$ by $\mathrm{asc}(w)$. An index $i \in \{1, \ldots, n-1\}$ is a *double ascent* if $w_{i-1} < w_i < w_{i+1}$. Analogously a *double descent* is an index $i \in \{1, \ldots, n-1\}$ with $w_{i-1} > w_i > w_{i+1}$.

Now we can describe the local $h$-vector of the edgewise subdivision, proved by Athanasiadis using methods of enumerative combinatorics.

**Theorem 2.33.** [Ath16a, Theorem 1.1] *The local $h$-polynomial of the $r$-th edgewise subdivision $\mathrm{esd}_r(2^V)$ of the $(n-1)$-dimensional simplex can be expressed as*

$$\ell_V(\mathrm{esd}_r(2^V), x) = E_r(x + x^2 + \cdots + x^{r-1})^n = \sum_{w \in S(n,r)} x^{\mathrm{asc}(w)}.$$

As a corollary of this theorem, Athanasiadis also showed the $\gamma$-nonnegativity of the local $h$-polynomial [Ath16a, Corollary 1.2], again by considering a suitable version of the valley-hopping method, similar as in the case for barycentric subdivisions.

**Corollary 2.34.** [Ath16a] *For all positive integers $n, r$ we have*

$$\ell_V(\mathrm{esd}_r(2^V), x) = \sum_{i=0}^{\lfloor \frac{n}{2} \rfloor} \xi_{n,r,i} x^i (1+x)^{n-2i},$$

*where $\xi_{n,r,i}$ is the number of sequences $w \in S(n, r)$ with exactly $i$ ascents fulfilling the following property: for every double ascent $k$ of $w$ there exists a double descent $l > k$ such that $w_k = w_l$ and $w_k \leq w_j$ for all $k < j < l$.*

The even stronger result, that the local $h$-polynomial of the $r$-th edgewise subdivision is real-rooted was proven by Zhang [Zha19, Theorem 1.1] by proving the interlacing property for a related family of polynomials, and independently by Leander [Lea16, Corollary 2.5] by finding a suitable set of compatible polynomials.

**Theorem 2.35.** [Zha19, Lea16] *The local h-polynomial of the r-th edgewise subdivision of the simplex* $\ell_V(\mathrm{esd}_r(2^V), x)$ *has only real zeros.*

In order to generalize Theorem 2.27, Athanasiadis [Ath14] introduced the so-called *r-colored barycentric subdivision*, which is the $r$-th edgewise subdivision of the barycentric subdivision. Since the $r$-th edgewise subdivision and the barycentric subdivision are uniform triangulations, so is the $r$-colored barycentric subdivison.

Using the Recursion (2.6) for general uniform triangulations, Athanasiadis gave a combinatorial interpretation of the $h$-vector transformation of the $r$-colored barycentric subdivision.

**Theorem 2.36.** [Ath20b, Proposition 4.7] *Let $\Delta$ be an $(n-1)$-dimensional simplicial complex. Then*

$$h_j(\mathrm{esd}_r(\mathrm{sd}(\Delta))) = \sum_{k=0}^{n} p_{\mathcal{F}}(n,k,j) h_k(\Delta),$$

*where $p_{\mathcal{F}}(n,k,j)$ is equal to the number of $r$-colored permutations $w \in \mathbb{Z}_r \wr \mathfrak{S}_{n+1}$ whose first coordinate has color zero and which have $j$ descents and last coordinate with color zero and equal to $n+1-k$.*

As a generalization of the description of the local $h$-vector for the barycentric subdivision in Theorem 2.27, Athanasiadis proved the following result for the local $h$-vector of the $r$-colored barycentric subdivision of the simplex. By setting $r=1$ one recovers Theorem 2.27.

**Theorem 2.37.** [Ath14, Theorem 1.2] *The local h-polynomial of the r-colored barycentric subdivision $\mathrm{esd}_r(\mathrm{sd}(2^V))$ of the $(n-1)$-dimensional simplex can be expressed as*

$$\ell_V(\mathrm{esd}_r(\mathrm{sd}(2^V)), x) = \sum_{w \in (D_n^r)^b} x^{\frac{\mathrm{fexc}(w)}{r}},$$

*where $(D_n^r)^b$ is the set of balanced derangements in $\mathbb{Z}_r \wr \mathfrak{S}_n$ and $\mathrm{fexc}(w)$ is the flag excedence of $w \in Z_r \wr \mathfrak{S}_n$.*

Furthermore, the local $h$-polynomial is $\gamma$-positive, which was shown by Athanasiadis [Ath14, Theorem 1.2], using that the two polynomials arising in the symmetric decomposition of the local $h$-polynomial are $\gamma$-nonnegative. [Ath14, Theorem 1.3]. Again, this theorem specializes to Theorem 2.28 for $r=1$.

**Theorem 2.38.** [Ath14] *The local $h$-polynomial of the $r$-colored barycentric subdivision can be expressed as*

$$\ell_V(\mathrm{esd}_r(\mathrm{sd}(2^V)), x) = \sum_{i=1}^{\lfloor \frac{n}{2} \rfloor} \xi^+_{n,r,i} x^i (1 + x)^{n-2i},$$

*where $\xi^+_{n,r,i}$ stands for the number of elements in $\mathbb{Z}_r \wr \mathfrak{S}_n$ with $i$ descending runs, no descending run of size one and last coordinate of color zero.*

Assuming, that $d_{n,r}(x) = d^+_{n,r}(x) + d^-_{n,r}(x)$ is the symmetric decomposition of the $n$-th $r$-colored derangement polynomial, it can be shown that $\ell_V(\mathrm{esd}_r(\mathrm{sd}(2^V)), x) = d^+_{n,r}(x)$. As discussed in Example 1.64, the polynomial $d_{n,r}(x)$ has a real-rooted symmetric decomposition, so in particular the local $h$-polynomial of the $r$-th colored barycentric subdivision is real-rooted.

### 2.4.3 The interval subdivision

The third example of a subdivision we describe is the interval subdivision, see Definition 2.7. Walker [Wal88] introduced this simplicial complex and proved that $\mathrm{Int}(\Delta)$ is a subdivision of $\Delta$ [Wal88, Theorem 6.1(a)].

**Example 2.39.** Figure 2.5 shows the interval subdivision of the 2-simplex, with some labeled vertices.

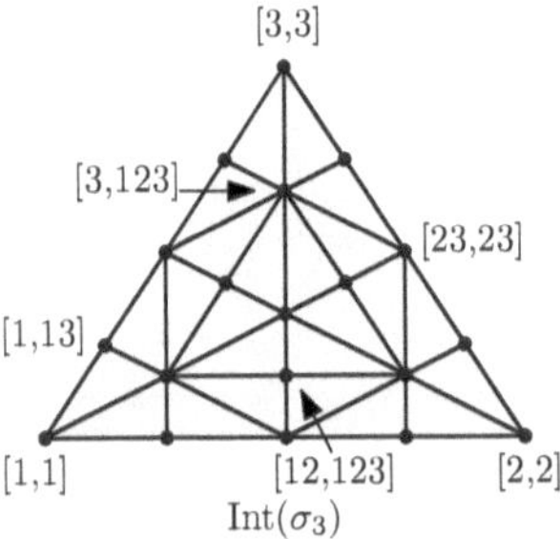


Figure 2.5: The interval subdivision of the 2-simplex

One can show, e.g., see [Ath16b, Remark 4.5], that the $f$-vector of this subdivision coincides with the $f$-vector of the 2-colored subdivision, discussed for the general case in the previous section.

In [AN20, Theorem 2.2], Anwar and Nazir gave an explicit description of the $f$-vector of the interval subdivision of an arbitrary complex $\Delta$, by counting the number of intervals of a fixed length and with a fixed top element:

**Theorem 2.40.** [AN20] *Let $\Delta$ be an $(n-1)$-dimensional simplicial complex. Then*

$$f_k(\mathrm{Int}(\Delta)) = \sum_{l=0}^{n}\sum_{i=0}^{k}(-1)^i\binom{k}{i}[(2+2k-2i)^l - (1+2k-2i)^l]f_{l-1}(\Delta)$$

$$= \sum_{l=0}^{n}\sum_{i=0}^{l}\binom{k}{i}k!S(j,k)[2^l - 2^i]f_{l-1}(\Delta)$$

*for $0 \le k \le n-1$ and $f_{-1}(\mathrm{Int}(\Delta)) = f_{-1}(\Delta) = 1$.*

To state the combinatorial interpretation of the $h$-vector transformation, we need to introduce some notation for the so-called *signed permutations*, denoted by $B_n$. Let $B_n$ be the group consisting of all the bijections $\sigma : [-n, n] \to [-n, n]$ such that $\sigma_{-1} = -\sigma_1$ for all $i \in [-n, n]$. An element $\sigma \in B_n$ is completely determined by $\sigma_1, \ldots, \sigma_n$ and we will write $\sigma = \sigma_1 \ldots \sigma_n$.

Analog to the permutation statistics in Section 1.2, we also define *descents* for the elements of $B_n$ to be indices $i \in [n]$ such that $\sigma_i > \sigma_{i+1}$. We also define *excedances* and *derangements* analogously.

Let $B_{n,j} = \{\sigma \in B_n : \sigma_1 = j\}$ and define the *$j$-Eulerian polynomial of type $B$* as

$$B_{n,j}(x) := \sum_{\sigma \in B_{n,j}} x^{\mathrm{des}(\sigma)} = \sum_{k=0}^{d} B(d,j,k)x^k,$$

where $B(d,j,k) = |\{\sigma \in B_{n,j} : \mathrm{des}(\sigma) = k\}|$.

Let $B_n^+ = \{\sigma \in B_n : \sigma_n > 0\}$ and $B_{n,j}^+ = \{\sigma \in B_n^+ : \sigma_1 = j\}$, and define analogously $B_n^-$ and $B_{n,j}^-$. Now we define the *$j$-Eulerian polynomial of type $B^+$* as

$$B_{n,j}^+(x) = \sum_{\sigma \in B_{n,j}^+} x^{\mathrm{des}(\sigma)} = \sum_{k=0}^{d-1} B^+(n,j,k)x^k,$$

where $B^+(n,j,k) = |\{\sigma \in B_{n,j}^+ : \mathrm{des}(\sigma) = k\}|$.

Now we can give an interpretation of the $h$-vector transformation proven by Anwar and Nazir [AN20, Theorem 3.1], similar to the one for the barycentric subdivision in Theorem 2.24.

**Theorem 2.41.** [AN20] *Let $\Delta$ be an $(n-1)$-dimensional simplicial complex. Then*

$$h_j(\mathrm{Int}(\Delta)) = \sum_{i=0}^{n} B^+(n+1, j+1, i)h_i(\Delta)$$

*for all $0 \le j \le n$.*

Using the theory of compatible polynomials, they show that the $h$-polynomials of interval subdivisions are real-rooted.

**Theorem 2.42.** [AN20, Theorem 4.1] *Let $\Delta$ be an $(n-1)$-dimensional simplicial complex such that the $h$-vector $h(\Delta) = (h_0(\Delta), \ldots, h_n(\Delta))$ is nonnegative. Then the $h$-polynomial*

$$h(\mathrm{Int}(\Delta), x) = \sum_{i=0}^{n} h_i(\mathrm{Int}(\Delta))x^i$$

*has only real roots.*

*In particular, the $h$-polynomial $h(Int(\Delta), x)$ is log-concave and unimodal.*

Athanasiadis and Savvidou [AS13, Corollary 1.2, Theorem 1.3, Proposition 4.1] proved, that the local $h$-polynomial of the interval subdivision is symmetric and unimodal and $\gamma$-nonnegative.

**Theorem 2.43.** [AS13] *The local $h$-polynomial of the interval subdivision is given by*

$$\ell_V(\mathrm{Int}(2^V), x) = \sum_{\sigma \in D_n \cap B_n^*} x^{\mathrm{exc}(\sigma)},$$

*where $B_n^* = \{\sigma \in B_n : \sigma_{m_\sigma} > 0\}$ and $m_\sigma$ is the minimal element in $\{\sigma_1, \ldots, \sigma_n\}$. It is symmetric, unimodal and $\gamma$-nonnegative.*

In [AN18, Remark 5.4], Anwar and Nazir conclude the $\gamma$-nonnegativity by interpreting the difference of $h(\mathrm{Int}(\sigma_n), x)$ and $h(\partial(\mathrm{Int}(\sigma_n)), x)$, and using a description of the local $h$-vector in terms of these differences given in [JKMS19, Theorem 4.4].

Furthermore, Anwar and Nazir proved that the $\gamma$-vector of the $h$-polynomial of the interval subdivision fulfills the Frankl-Füredi-Kalai inequalities, i.e., is the $f$-vector of a balanced simplicial complex.

**Theorem 2.44.** [AN18, Theorem 1.3] *If $\Delta$ is a simplicial complex with nonnegative and symmetric $h$-vector $h(\Delta)$, then the $\gamma$-vector of the interval subdivision of $\Delta$ is the $f$-vector of a balanced simplicial complex.*

# 3 Combinatorics of the antiprism triangulation

The following chapter is based on the paper "Combinatorics of antiprism triangulations" [ABJK20].

As seen in the previous section, (uniform) triangulations and subdivisions of simplicial complexes can admit nice enumerative and algebraic properties. The typical example is the barycentric subdivision (see Chapter 2.4.1). For example, it has been shown that the transformation of the $h$-vector of a simplicial complex under the barycentric subdivision can be described explicitly in combinatorial terms, see Theorem 2.24, and the $h$-polynomial of the barycentric subdivision of a simplicial complex $\Delta$ has only real roots (implying log-concave and unimodal coefficients) for every complex $\Delta$ with nonnegative $h$-vector, see Theorem 2.26.

A similar, but combinatorially more intricate and much less studied than barycentric subdivision, way to subdivide $\Delta$ is provided by the *antiprism triangulation*, denoted by $\mathrm{sd}_A(\Delta)$.

The antiprism triangulation was introduced by Izmestiev and Joswig [IJ03] as a technical device in their effort to understand combinatorially branched coverings of manifolds, and arose independently and was studied under the name *chromatic subdivision* in computer science (specifically, in theoretical distributed computing); see [Koz12] and references therein.

We show that the antiprism triangulation too has very interesting enumerative and algebraic properties and that its study can lead to challenging combinatorial problems. Our main motivation comes from the following conjectural analogue of Theorem 2.26, the main result of [BW08].

**Conjecture 3.1.** *The polynomial $h(\mathrm{sd}_A(\Delta), x)$ is real-rooted for every simplicial complex $\Delta$ with nonnegative $h$-vector.*

Although we are unable to fully settle this conjecture, we reduce it to an interlacing relation between the members of two concrete infinite sequences of polynomials (see Conjecture 3.19), given the following important special case of the conjecture and Theorem 2.21.

**Theorem 3.2.** *The polynomial $h(\mathrm{sd}_A(\sigma_n), x)$ is real-rooted and has a nonnegative, real-rooted and interlacing symmetric decomposition with respect to $n-1$ for every positive integer $n$.*

In this Chapter we study combinatorial properties of the antiprism triangulation in the spirit of the results seen in the previous chapter for other subdivisions.

In the first section, we describe the antiprism construction which can be used to define the antiprism triangulation. We prove recurrences for the $h$- and local $h$-polynomial of antiprisms over uniform triangulations, which will be useful for the special case of antiprism triangulations.

In the second part, we define the antiprism triangulation in several ways, and give a description of the faces by combinatorial objects.

In the following, we first consider the special case of the antiprism triangulation of the simplex, and prove combinatorial interpretations of its $h$- and local $h$-vector and the real-rootedness of its $h$-polynomial and related polynomials.

Finally, we consider the antiprism triangulation of an arbitrary simplicial complex, and its $h$-vector transformation.

## 3.1 The antiprism construction

This section reviews the antiprism construction and studies its face enumeration, in the framework of uniform triangulations [Ath20b]. The results will be applied in the next sections in the case of uniform triangulations, but may be of independent interest too.

**Definition 3.3.** Let $V = \{v_1, v_2, \ldots, v_n\}$ be an $n$-element set and $\Delta$ be a triangulation of the boundary complex of the simplex $2^V$. Pick an $n$-element set $U = \{u_1, u_2, \ldots, u_n\}$ which is disjoint from the vertex set of $\Delta$ and denote by $\Gamma_{\mathcal{A}}(\Delta)$ the collection of faces of $\Delta$ together with all sets of the form $E \cup G$, where $E = \{u_i \: : \: i \in I\}$ is a nonempty face of the simplex $2^U$ for some $\varnothing \subsetneq I \subseteq [n]$ and $G$ is a face of the restriction of $\Delta$ to the face $F = \{v_i \: : \: i \in [n] \smallsetminus I\}$ of $\partial(2^V)$ which is complementary to $E$. The collection $\Gamma_{\mathcal{A}}(\Delta)$ is a simplicial complex which contains $2^U$ and $\Delta$ as subcomplexes and we call it the *antiprism over* $\Delta$.

**Example 3.4.** When $\Delta = \partial(2^V)$ is the trivial triangulation, the antiprism $\Gamma_{\mathcal{A}}(\partial(2^V))$ is combinatorially isomorphic to the Schlegel diagram [Zie95, Section 5.2] of the $n$-dimensional cross-polytope behind any of its facets. Figure 3.1 shows the antiprism over the boundary of the 2-dimensional simplex.

For general $\Delta$, the antiprism $\Gamma_{\mathcal{A}}(\Delta)$ is a triangulation of $\Gamma_{\mathcal{A}}(\partial(2^V))$: the carrier of a face $E \cup G$, as above, is the union of $E$ with the carrier of $G$, the latter considered as a face of the triangulation $\Delta$ of $\partial(2^V)$. Since $\Gamma_{\mathcal{A}}(\partial(2^V))$ triangulates the simplex $2^V$, $\Gamma_{\mathcal{A}}(\Delta)$ is a triangulation of $2^V$ as well with boundary equal to $\Delta$.

**Remark 3.5.** Given a triangulation $\Gamma$ of the $(n-1)$-dimensional simplex $2^V$, an analogous procedure defines a triangulation, say $\Delta_{\mathcal{A}}(\Gamma)$, of the $(n-1)$-dimensional sphere which contains $2^U$ and $\Gamma$ as subcomplexes and which we may call the *antiprism over* $\Gamma$. This construction was employed in [Ath12, Section 4], in order to relate the $\gamma$-vector of a flag triangulation of the sphere to the local $\gamma$-vector of a flag triangulation of the simplex, and in [Ath20a, Section 4], in order to interpret geometrically binomial Eulerian polynomials (see Example 3.9) and certain analogues

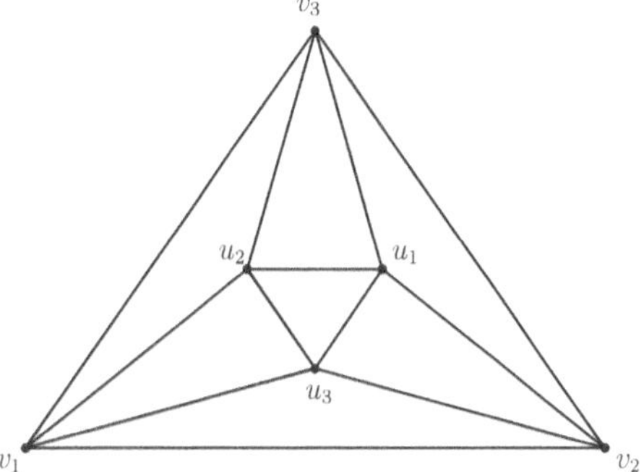



Figure 3.1: $\Gamma_{\mathcal{A}}(\partial(\sigma_3))$

for $r$-colored permutations. The connection between the two constructions is that
$\Delta_{\mathcal{A}}(\Gamma) = \Gamma \cup \Gamma_{\mathcal{A}}(\partial\Gamma)$. $\qquad\square$

The following statement is closely related to [Ath20a, Proposition 4.1].

**Proposition 3.6.** *The simplicial complex* $\Gamma_{\mathcal{A}}(\Delta)$ *triangulates the* $(n-1)$*-dimensional simplex* $2^V$ *for every triangulation* $\Delta$ *of the boundary complex* $\partial(2^V)$. *Moreover,*

$$h(\Gamma_{\mathcal{A}}(\Delta), x) \;=\; \sum_{F \subsetneq V} x^{|F|} h(\Delta_F, 1/x).$$

*Proof.* We have already commented on the first sentence. For the second, using Proposition 1.15 and the definition of the $h$-polynomial we find that

$$
\begin{aligned}
x^n h(\Gamma_{\mathcal{A}}(\Delta), 1/x) \;=\; h^{\circ}(\Gamma_{\mathcal{A}}(\Delta), x) \;&=\; \sum_{G \in \Gamma_{\mathcal{A}}(\Delta)^{\circ}} x^{|G|}(1-x)^{n-|G|} \\
&=\; \sum_{\varnothing \neq E \subseteq U} \ \sum_{\substack{G \in \Gamma_{\mathcal{A}}(\Delta), \\ G \cap U = E}} x^{|G|}(1-x)^{n-|G|}.
\end{aligned}
$$

By definition of $\Gamma_{\mathcal{A}}(\Delta)$, the inner sum is equal to $x^{|E|} h(\Delta_F, x)$, where $F \subsetneq V$ is the face of $2^V$ which is complementary to $E$. Replacing $x$ by $1/x$ results in the proposed expression for $h(\Gamma_{\mathcal{A}}(\Delta), x)$ and the proof follows. $\qquad\square$

We now turn our attention to uniform triangulations of $\partial(2^V)$.

**Proposition 3.7.** *For every* $\mathcal{F}$*-uniform triangulation* $\Delta$ *of the boundary complex of an* $(n-1)$*-dimensional simplex* $2^V$:

$$h(\Gamma_A(\Delta), x) \;=\; \sum_{k=0}^{n-1} \binom{n}{k} x^k h_{\mathcal{F}}(\sigma_k, 1/x) \tag{3.1}$$

$$=\; \sum_{k=0}^{n-1} \binom{n}{k} \ell_{\mathcal{F}}(\sigma_k, x) \left( (1+x)^{n-k} - x^{n-k} \right), \tag{3.2}$$

$$\ell_V(\Gamma_A(\Delta), x) \;=\; \sum_{k=0}^{n-1} \binom{n}{k} \ell_{\mathcal{F}}(\sigma_k, x) \left( (1+x)^{n-k} - 1 - x^{n-k} \right), \tag{3.3}$$

$$h(\Gamma_A(\Delta), x) - h(\Delta, x) \;=\; \sum_{k=0}^{n-1} \binom{n}{k} \ell_{\mathcal{F}}(\sigma_k, x) \left( (1+x)^{n-k} - 1 - x - \cdots - x^{n-k} \right) \tag{3.4}$$

$$=\; \sum_{k=0}^{n-1} \binom{n}{k} h_{\mathcal{F}}(\sigma_k, x) \left( x^{n-k} - x(x-1)^{n-k-1} \right). \tag{3.5}$$

*In particular, if all restrictions of $\Delta$ to proper faces of $2^V$ are regular triangulations, then the polynomials $\ell_V(\Gamma_A(\Delta), x)$ and $h(\Gamma_A(\Delta), x) - h(\Delta, x)$ are unimodal and $h(\Gamma_A(\Delta), x)$ is alternatingly increasing with respect to $n-1$.*

*Proof.* Equation (3.1) follows directly from Proposition 3.6. To deduce Equation (3.2) from that, we use (2.1) to express $h_{\mathcal{F}}(\sigma_k, 1/x)$ in terms of local $h$-polynomials, apply the symmetry property of the latter and change the order of summation to obtain

$$h(\Gamma_A(\Delta), x) \;=\; \sum_{k=0}^{n-1} \binom{n}{k} x^k \sum_{j=0}^{k} \binom{k}{j} \ell_{\mathcal{F}}(\sigma_j, 1/x) \;=\; \sum_{k=0}^{n-1} \binom{n}{k} \sum_{j=0}^{k} x^{k-j} \binom{k}{j} \ell_{\mathcal{F}}(\sigma_j, x)$$

$$=\; \sum_{j=0}^{n-1} \ell_{\mathcal{F}}(\sigma_j, x) \sum_{k=j}^{n-1} \binom{n}{k} \binom{k}{j} x^{k-j} \;=\; \sum_{j=0}^{n-1} \binom{n}{j} \ell_{\mathcal{F}}(\sigma_j, x) \sum_{k=j}^{n-1} \binom{n-j}{n-k} x^{k-j}$$

$$=\; \sum_{j=0}^{n-1} \binom{n}{j} \ell_{\mathcal{F}}(\sigma_j, x) \left( (1+x)^{n-j} - x^{n-j} \right).$$

For the fourth and fifth step we have used the identity $\binom{n}{k}\binom{k}{j} = \binom{n}{j}\binom{n-j}{n-k}$ and the binomial theorem, respectively.

Alternatively, Equation (3.2) follows from an application of Stanley's locality formula (2.2) to $\Gamma_A(\Delta)$, considered as a triangulation of the antiprism $\Gamma_A(\partial(2^V))$ over the boundary complex of $2^V$. Equation (3.3) follows when combining (3.2) with

$$h(\Gamma_A(\Delta), x) \;=\; \ell_V(\Gamma_A(\Delta), x) + \sum_{k=0}^{n-1} \binom{n}{k} \ell_{\mathcal{F}}(\sigma_k, x), \tag{3.6}$$

the latter being (2.1) applied to $\Gamma_A(\Delta)$. Equation (3.4) follows from (3.2) and

$$h(\Delta, x) \;=\; \sum_{k=0}^{n-1} \binom{n}{k} \ell_{\mathcal{F}}(\sigma_k, x)(1 + x + x^2 + \cdots + x^{n-k-1}), \tag{3.7}$$

which is also a consequence of [Sta92, Theorem 3.2]; see [JKMS19, Equation (4.2)]. Equation (3.5) follows from (3.4) by expressing $\ell_{\mathcal{F}}(\sigma_k, x)$ in terms of the $h$-polynomials $h_{\mathcal{F}}(\sigma_j, x)$, changing the order of summation and computing the inner sum, just as in the proof of Equation (3.2); we leave the details of this computation to the interested reader.

For the last statement we note that, by the regularity assumption, $\ell_{\mathcal{F}}(\sigma_k, x)$ is (symmetric with center of symmetry $k/2$ and) unimodal for $0 \leq k < n$. As a result, Equations (3.3) and (3.4) imply the unimodality of $\ell_V(\Gamma_A(\Delta), x)$ and $h(\Gamma_A(\Delta), x) - h(\Delta, x)$, respectively, and Equations (3.4) and (3.7) imply that the symmetric decomposition

$$h(\Gamma_A(\Delta), x) \; = \; h(\Delta, x) + (h(\Gamma_A(\Delta), x) - h(\Delta, x))$$

of $h(\Gamma_A(\Delta), x)$ with respect to $n - 1$ is nonnegative and unimodal. $\qquad\square$

**Remark 3.8.** Let $\Delta$ be as in Proposition 3.7. Since coning a simplicial complex does not affect the $h$-polynomial, the right-hand side of (3.7) is also an expression for $h(u * \Delta, x)$. The formula

$$h(u * \Delta, x) \; = \; \sum_{k=0}^{n-1} \binom{n}{k} h_{\mathcal{F}}(\sigma_k, x)(x - 1)^{n-k-1} \tag{3.8}$$

can be derived from that by expressing $\ell_{\mathcal{F}}(\sigma_k, x)$ in terms of the $h$-polynomials $h_{\mathcal{F}}(\sigma_j, x)$, changing the order of summation and computing the inner sum, just as in the proof of Equations (3.2) and (3.5) or, alternatively, by adapting the argument in the proof of Proposition 3.6. When $\Delta$ is the barycentric subdivision of $\partial\sigma_n$, this yields the recursion

$$A_n(x) \; = \; \sum_{k=0}^{n-1} \binom{n}{k} A_k(x)(x - 1)^{n-k-1}$$

for the Eulerian polynomial $A_n(x)$, valid for $n \geq 1$. This appears as [Foa10, Equation (2.7)].

**Example 3.9.** Suppose again that $\Delta$ is the barycentric subdivision of $\partial\sigma_n$. Then, Equation (3.1) yields that

$$h(\Gamma_A(\Delta), x) \; = \; \sum_{k=0}^{n-1} \binom{n}{k} x^k A_k(1/x) \; = \; 1 + x\sum_{k=1}^{n-1} \binom{n}{k} A_k(x) \; = \; \widetilde{A}_n(x) - xA_n(x)$$

and $h(\Gamma_A(\Delta), x) - h_{\mathcal{F}}(\partial\sigma_n) = \widetilde{A}_n(x) - (1 + x)A_n(x)$, where

$$\widetilde{A}_n(x) \; := \; 1 + x\sum_{k=1}^{n} \binom{n}{k} A_k(x)$$

is the $n$-th binomial Eulerian polynomial studied, for instance, in [Ath20a, SW20]. From Equation (3.6) we compute further that $\ell_V(\Gamma_A(\Delta), x) = \widetilde{A}_n(x) - (1+x)A_n(x) - d_n(x)$, where $d_n(x) = \ell_{\mathcal{F}}(\sigma_n, x)$ is the $n$-th derangement polynomial (see Section 1.2.1).

Therefore, by Proposition 3.7, $\widetilde{A}_n(x) - xA_n(x)$ is alternatingly increasing with respect to $n - 1$ and $\widetilde{A}_n(x) - (1 + x)A_n(x)$ is symmetric and unimodal. $\qquad\square$

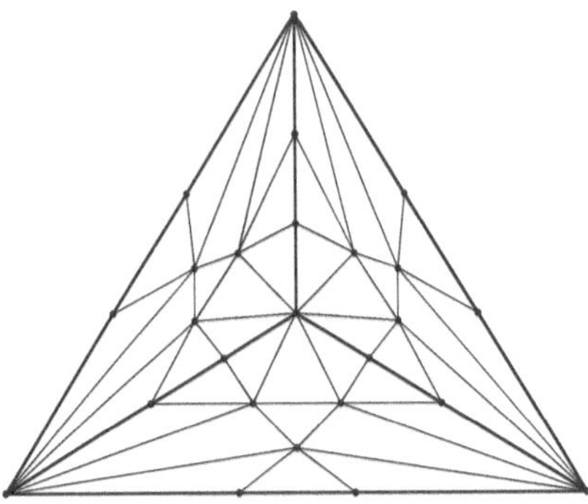

Figure 3.2: Antiprism triangulation of the cone over the boundary of the 2-simplex

## 3.2 The antiprism triangulation

This section describes combinatorially and geometrically the antiprism triangulation of a simplicial complex. For more information we also refer to [IJ03, Apendix A.1] and [Koz12], where these descriptions are given in variant forms.

The antiprism triangulation can be defined similarly as the barycentric subdivision in Definition 2.5, if the nonempty faces of $\Delta$ are replaced by pointed faces and coning is replaced by the antiprism construction of Section 3.1.

**Definition 3.10.** A *pointed subset* of a set $V$ is any pair $(S, v)$ such that $v \in S \subseteq V$. Similarly, a *pointed face* of a simplicial complex $\Delta$ is any pair $(F, v)$ such that $F \in \Delta$ is a face and $v \in F$ is a chosen vertex.

**Definition 3.11.** Let $\Delta$ be a simplicial complex. We denote by $\mathcal{G}_A(\Delta)$ the (simple, undirected) graph on the node set of pointed faces of $\Delta$ for which two distinct pointed faces $(F, v)$ and $(F', v')$ are adjacent if

- $F = F'$, or
- $F \subsetneq F'$ and $v' \in (F' \smallsetminus F)$, or
- $F' \subsetneq F$ and $v \in (F \smallsetminus F')$.

The *antiprism triangulation* of $\Delta$, denoted by $\mathrm{sd}_A(\Delta)$, is the abstract simplicial complex defined as the clique complex of $\mathcal{G}_A(\Delta)$.

Examples of antiprism triangulations are shown in Figures 3.2 and 3.3.

Alternatively, we can define this subdivision in the following two equivalent ways:

The antiprism subdivision $\mathrm{sd}_A(\sigma_n)$ is the subdivision of the $(n-1)$-dimensional simplex with vertex set the pointed faces of $\sigma_n$, and

1. the facets of $\mathrm{sd}_A(\sigma_n)$ are the sets of pointed faces

$$\{(F_1, 1), (F_2, 2), \ldots, (F_n, n)\}$$

   satisfying the following two conditions:

   (C1) for all $i, j \in [n]$ one has either $F_i \subseteq F_j$ or $F_j \subseteq F_i$,

   (C2) for all $i, j \in [n]$, if $i \in F_j$, then $F_i \subseteq F_j$.

2. a set $\{(F_0, v_0), \ldots, (F_k, v_k)\}$ is a $k$-face if and only if all the $v_i$ are distinct and the following two properties hold:

   (C1*) $F_0 \subseteq F_1 \subseteq \cdots \subseteq F_k$ is a flag in $\Delta$ (with repetitions allowed),

   (C2*) If $F_i \subsetneq F_j$, then $v_j \notin F_i$.

Alternativley, the faces of $\mathrm{sd}_A(\Delta)$ can also be described explicitly in other combinatorial terms [Koz12, Section 2].

**Definition 3.12.** Given a set $S$, an *ordered set partition* (or simply, ordered partition) of $S$ is any sequence of nonempty, pairwise disjoint sets (called *blocks*) whose union is equal to $S$. A *multi-pointed ordered partition* of $S$ is defined as a pair $(\pi, \tau)$, where $\pi = (B_1, B_2, \ldots, B_m)$ and $\tau = (C_1, C_2, \ldots, C_m)$ are ordered partitions of $S$ and of a subset of $S$, respectively, with the same number of blocks, such that $C_i$ of a nonempty subset of $B_i$ for every $i \in [m]$. We can think of such a pair as an ordered partition of $S$, together with a choice of a nonempty subset for every block. The sum of the cardinalities of these subsets $C_i$ (total number of chosen elements) will be called the *weight* of $(\pi, \tau)$.

Then, the $(k-1)$-dimensional faces of $\mathrm{sd}_A(\Delta)$ are in one-to-one correspondence with the multi-pointed ordered partitions of faces of $\Delta$ of weight $k$. More specifically, the multi-pointed ordered partition $(\pi, \tau)$, with $\pi = (B_1, B_2, \ldots, B_m)$ and $\tau = (C_1, C_2, \ldots, C_m)$, corresponds to the face of $\mathrm{sd}_A(\Delta)$ with vertices the pointed faces $(F, v)$ of $\Delta$, where $F = B_1 \cup B_2 \cup \cdots \cup B_i$ for some $i \in [m]$ and $v \in C_i$. The faces of the antiprism triangulation of the simplex $2^V$ are the multi-pointed ordered partitions of subsets of $V$; they will be referred to as *multi-pointed partial ordered partitions* of $V$. Note that the facets of $\mathrm{sd}_A(\Delta)$ are in one-to-one correspondence with the ordered partitions of the facets of $\Delta$ (since all elements in the blocks should be chosen).

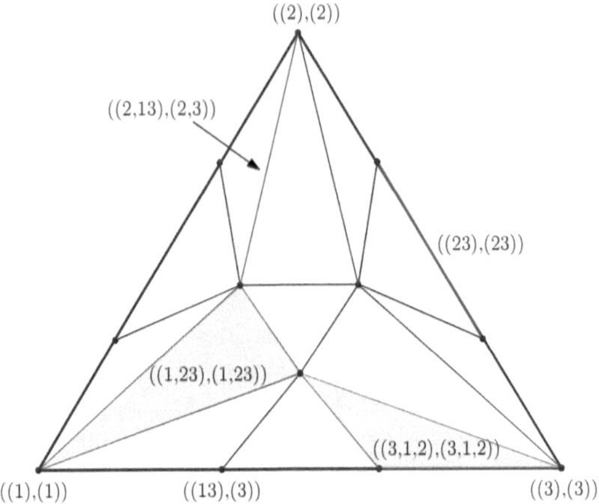


Figure 3.3: Antiprism triangulation of the 2-simplex

**Example 3.13.** Figure 3.3 shows the antiprism triangulation of the 2-simplex, including some faces labeled by multi-pointed ordered partitions.

The facet $((2,13),(2,13))$ for example corresponds to the set of vertices $\{(\{2\},2),(\{123\},1),(\{123\},3)\}$.

As was the case with barycentric subdivision (see Definition 2.5), $\mathrm{sd}_A(\Delta)$ can be constructed geometrically by applying the antiprism construction of Section 3.1 to its faces, starting from the edges and moving to faces of higher dimension in any order which respects inclusion. This process is slightly different from the one in [IJ03, Koz12] which uses crossing operations on the faces of $\Delta$ instead.

**Definition 3.14.** A *crossing operation* (also known as a *balanced stellar subdivision* [BM87]) on a face $F \in \Delta$ replaces $\mathrm{st}_\Delta(F)$ by the join of $\mathrm{lk}_\Delta(F)$ with the antiprism (as defined in Section 3.1) over $\partial(2^F)$.

Performing this operation starting from facets and moving to faces of lower dimension in any order which respects reverse inclusion yields a triangulation. Both approaches result in a triangulation which is combinatorially isomorphic to $\mathrm{sd}_A(\Delta)$. In Figure 3.4 and Figure 3.5 we show the construction of $\mathrm{sd}_A(\sigma_2)$ using the antiprism construction and the crossing operation respectively. Under the previously mentioned isomorphism, the carrier of a multi-pointed ordered partition of a face $F \in \Delta$ is equal to $F$. As a result, the interior faces of the antiprism triangulation of the simplex $2^F$ are in one-to-one correspondence with the multi-pointed ordered partitions of $F$.

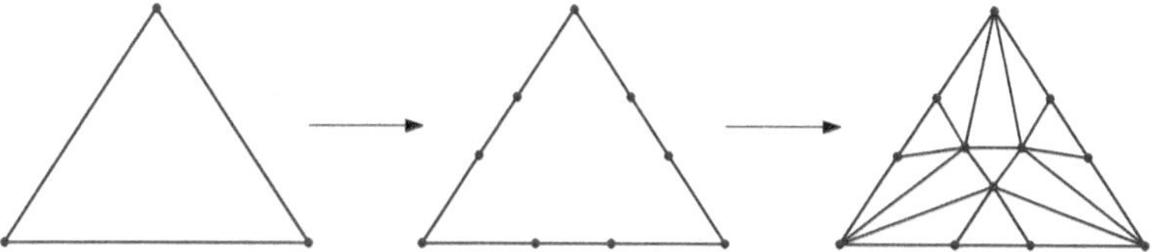

Figure 3.4: Triangulation of $\sigma_2$ via the antiprism operation

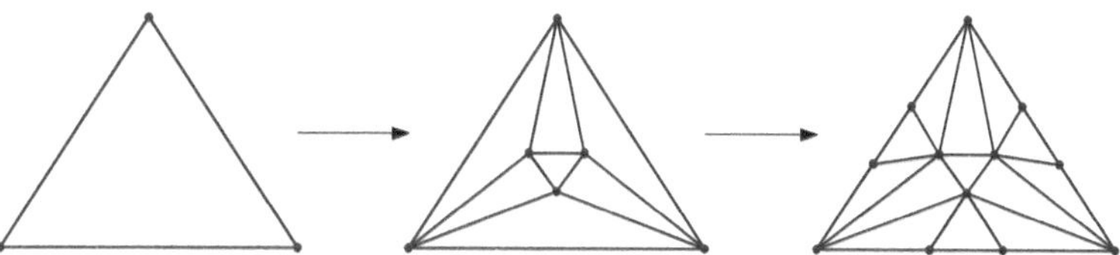

Figure 3.5: Triangulation of $\sigma_2$ via crossing operation

## 3.3 The antiprism triangulation of a simplex

As discussed in Section 3.2, the number of $(k-1)$-dimensional faces of the antiprism triangulation $\mathrm{sd}_A(\sigma_n)$ is equal to the number of multi-pointed partial ordered set partitions of $[n]$ of weight $k$. We now give a recurrence and combinatorial interpretations for the $h$-polynomial of $\mathrm{sd}_A(\sigma_n)$. For the first few values of $n$,

$$
h_A(\sigma_n, x) \;=\; \begin{cases}
1, & \text{if } n = 0 \\
1, & \text{if } n = 1 \\
1 + 2x, & \text{if } n = 2 \\
1 + 9x + 3x^2, & \text{if } n = 3 \\
1 + 28x + 42x^2 + 4x^3, & \text{if } n = 4 \\
1 + 75x + 310x^2 + 150x^3 + 5x^4, & \text{if } n = 5 \\
1 + 186x + 1725x^2 + 2300x^3 + 465x^4 + 6x^5, & \text{if } n = 6 \\
1 + 441x + 8211x^2 + 23625x^3 + 13685x^4 + 1323x^5 + 7x^6, & \text{if } n = 7.
\end{cases}
$$

We first need to introduce some more terminology.

**Definition 3.15.** Let $\varphi = (\pi, \tau)$ be a multi-pointed partial ordered set partition of $[n]$. Thus, $\pi = (B_1, B_2, \ldots, B_m)$ is an ordered partition of a subset $S$ of $[n]$ and

$\tau = (C_1, C_2, \ldots, C_m)$, where $C_i$ is a nonempty subset of $B_i$ for every $i \in [m]$. We will say that $\varphi$ is *proper* if $C_i$ is a proper subset of $B_i$ for every $i \in [m]$. We will use the same terminology with the adjective 'partial' dropped, when $S = [n]$.

**Proposition 3.16.** (a) *We have*

$$h_{\mathcal{A}}(\sigma_n, x) \;=\; \sum_{k=0}^{n-1} \binom{n}{k} x^k h_{\mathcal{A}}(\sigma_k, 1/x) \tag{3.9}$$

*for every positive integer $n$.*

(b) *The coefficient of $x^k$ in $h_{\mathcal{A}}(\sigma_n, x)$ is equal to:*

- *the number of proper multi-pointed partial ordered set partitions of $[n]$ of weight $k$,*

- *the number of ways to choose a subset $S \subseteq [n]$ and an ordered set partition $\pi$ of $S$ and to color $k$ elements of $S$ black and the remaining elements white, so that no block of $\pi$ is monochromatic,*

- *the number of ordered set partitions $\pi = (B_1, B_2, \ldots, B_m)$ of $[n]$ for which the union $\bigcup_{i=1}^{\lfloor m/2 \rfloor} B_i$ has exactly $k$ elements,*

- $\binom{n}{k}$ *times the number of permutations in $\mathfrak{S}_n$ with excedance set equal to $[k]$,*

- *the explicit expression*

$$\binom{n}{k} \sum_{j=1}^{k+1} (-1)^{k+1-j} j! S(k+1, j) j^{n-k-1},$$

*where $S(n, k)$ are the Stirling numbers of the second kind.*

*Proof.* Part (a) follows from Proposition 3.7, as a special case of Equation (3.1). For part (b), we first note that from Equation (3.3) of the same proposition and Equation (2.1) we get

$$\ell_{\mathcal{A}}(\sigma_n, x) \;=\; \sum_{m=0}^{n-1} \binom{n}{m} \ell_{\mathcal{A}}(\sigma_m, x) \left( (1+x)^{n-m} - 1 - x^{n-m} \right)$$

for $n \geq 1$ and

$$h_{\mathcal{A}}(\sigma_n, x) \;=\; \sum_{m=0}^{n} \binom{n}{m} \ell_{\mathcal{A}}(\sigma_m, x),$$

respectively. By induction on $n$, the former equality implies that the coefficient of $x^k$ in $\ell_{\mathcal{A}}(\sigma_n, x)$ is equal to the number of proper multi-pointed ordered set partitions of $[n]$ of weight $k$. This and the latter equation yield the first interpretation of $h_{\mathcal{A}}(\sigma_n, x)$ claimed in part (b). The second interpretation is a restatement of the first (where black elements correspond to the chosen elements in the blocks of the multi-pointed partition).

(Alternatively, one can prove the second interpretation directly from the explicit expression provided in Proposition 3.27 for the special case $k = 0$ via an involution.)

The third interpretation can be deduced from the first as follows. Let $Q(n, k)$ denote the collection of proper multi-pointed partial ordered partitions of $[n]$ of weight $k$. Each element of $Q(n, k)$ is a triple consisting of a subset $S \subseteq [n]$, an ordered partition $\pi = (B_1, B_2, \ldots, B_r)$ of $S$ and a choice of nonempty proper subset $C_i$ of $B_i$ for every $i \in [r]$, such that the union $\cup_{i=1}^{r} C_i$ has cardinality $k$. From such a triple one can define an ordered partition of $[n]$ by listing the blocks $C_1, \ldots, C_r, B_1 \setminus C_1, \ldots, B_r \setminus C_r$ in this order and, if nonempty, adding $[n] \setminus S$ at the end as the last block. It is straightforward to verify that the resulting map is a bijection from $Q(n, k)$ to the collection of ordered partitions of $[n]$ described in the third proposed interpretation.

For the last two claimed interpretations, let us denote by $c(n, k)$ the number of permutations in $\mathfrak{S}_n$ with excedance set equal to $[k]$, for $k \in \{0, 1, \ldots, n\}$. Then, $c(n, n) = 0$ and, as a consequence of Lemma 2.2 and Theorem 2.5 in [ES00] (see also Section 3 of this reference), $c(n, k) = c(n, n - k - 1)$ and

$$c(n, k) \; = \; 1 + \sum_{m=1}^{k} \binom{k+1}{m} c(n - k - 1 + m, m).$$

for $k \in \{0, 1, \ldots, n - 1\}$. In view of $c(n, k) = c(n, n - k - 1)$, the latter equality can be rewritten as

$$c(n, k) \; = \; 1 + \sum_{m=1}^{n-k-1} \binom{n-k}{m} c(k + m, m). \tag{3.10}$$

On the other hand, writing $h_A(\sigma_n, x) = \sum_{k=0}^{n} p_A(n, k) x^k$ for $n \in \mathbb{N}$, the recursion of part (a) gives that

$$p_A(n, k) \; = \; \sum_{m=k}^{n-1} \binom{n}{m} p_A(m, m - k)$$

for $k \in \{0, 1, \ldots, n - 1\}$. Setting

$$p_A(n, k) \; = \; \binom{n}{k} \bar{p}_A(n, k),$$

the last recursion can be rewritten as

$$\binom{n}{k} \bar{p}_A(n, k) \; = \; \sum_{m=k}^{n-1} \binom{n}{m} \binom{m}{k} \bar{p}_A(m, m - k) \quad \text{i.e.,}$$

$$\bar{p}_A(n, k) \; = \; \sum_{m=k}^{n-1} \binom{n-k}{m-k} \bar{p}_A(m, m - k) \; = \; \sum_{m=0}^{n-k-1} \binom{n-k}{m} \bar{p}_A(k + m, m).$$

Comparing this recursion to (3.10) we get that $\bar{p}_A(n, k) = c(n, k)$ for all $n, k$. This proves the next to last interpretation, claimed in part (b). The last interpretation follows from this and the explicit formula for $c(n, k)$ obtainedin [ES00, Proposition 6.5]. $\qquad\square$

**Example 3.17.** We saw before, that for $n = 3$ we get $h(\mathrm{sd}_A(\sigma_3), x) = 1 + 9x + 3x^2$. These three coefficients can be interpreted in the following way. Regarding the second interpretation in Proposition 3.16(b), we have three ways to choose a subset and a partition of this subset with two colored elements without any monochromatic block, namely $(1, 2, 3), (1, 2, 3), (1, 2, 3)$. For one colored element, we get nine possible choices: $(1, 2), (1, 2), (1, 3), (1, 3), (2, 3), (2, 3), (1, 2, 3), (1, 2, 3), (1, 2, 3)$. For no colored element, we can only choose the emptyset, giving one choice.

For the ordered set partition in the third interpretation we get the three partitions $(12, 3), (13, 2), (23, 1)$ with two elements in the first half of the blocks. We have nine partitions $(1, 23), (2, 13), (3, 12), (1, 2, 3), (1, 3, 2), (2, 1, 3), (2, 3, 1), (3, 1, 2), (3, 2, 1)$ with one element in the first half of the blocks, and we can associate the partition $(123)$ to the remaining coefficient.

In view of the fourth interpretation, we have one permutation with excedance set $[2]$, namely $(231)$. There are three permutations with excedance set $[1]$, which are $(312), (321), (213)$, and one permutation $(123)$ with an empty excedance set.

The following statement is the main result of this section.

**Theorem 3.18.** *The polynomial $h_A(\sigma_n, x)$ is real-rooted and interlaces $h_A(\sigma_{n+1}, x)$ for every $n \in \mathbb{N}$. Moreover, $h_A(\sigma_n, x)$ has a nonnegative, real-rooted and interlacing symmetric decomposition with respect to $n - 1$, for every positive integer $n$.*

*Proof.* We consider the polynomials

$$q_{n,r}(x) := \sum_{k=0}^{n} \binom{n}{k} x^{k+r} h_A(\sigma_{k+r}, 1/x),$$

shown in Table 3.1 for the first few values of $n, r \in \mathbb{N}$. By part (a) of Proposition 3.16 and the definition of $q_{n,r}(x)$ we have

$$q_{n,0}(x) = h_A(\sigma_n, x) + x^n h_A(\sigma_n, 1/x), \tag{3.11}$$
$$q_{0,r}(x) = x^r h_A(\sigma_r, 1/x) \tag{3.12}$$

for every positive integer $n$ and every $r \in \mathbb{N}$, respectively. We claim that

$$\mathfrak{Q}_n := (q_{n,0}(x), q_{n-1,1}(x), \ldots, q_{1,n-1}(x), q_{0,n}(x), q_{0,n+1}(x))$$

is an interlacing sequence of real-rooted polynomials for every $n \in \mathbb{N}$. In particular, selecting the first and last two terms, we have the interlacing sequence

$$(h_A(\sigma_n, x) + x^n h_A(\sigma_n, 1/x), x^n h_A(\sigma_n, 1/x), x^{n+1} h_A(\sigma_{n+1}, 1/x))$$

of real-rooted polynomials for every $n \in \mathbb{N}$. Before we prove the claim let us observe that, since $x^n h_A(\sigma_n, 1/x)$ and $x^{n+1} h_A(\sigma_{n+1}, 1/x)$ have degrees $n$ and $n + 1$, respectively, the statement that the former polynomial interlaces the latter is equivalent to the statement that $h_A(\sigma_n, x)$ interlaces $h_A(\sigma_{n+1}, x)$. Similarly, since $h_A(\sigma_n, x) +$

$x^n h_{\mathcal{A}}(\sigma_n, 1/x)$ is symmetric of degree $n$, the statement that this polynomial interlaces $x^{n+1} h_{\mathcal{A}}(\sigma_{n+1}, 1/x)$ is equivalent to the statements that the same polynomial is interlaced by $x^n h_{\mathcal{A}}(\sigma_{n+1}, 1/x)$ and that it interlaces $h_{\mathcal{A}}(\sigma_{n+1}, x)$.

We now prove the claim by induction on $n$. This is true for $n = 0$, since $\mathcal{Q}_0 = (1, x)$. We assume that it holds for $n - 1 \in \mathbb{N}$. The standard recurrence for the binomial coefficients shows that $q_{n,r}(x) = q_{n-1,r}(x) + q_{n-1,r+1}(x)$ for every $r \in \mathbb{N}$. Writing this in the form

$$q_{n-r,r}(x) \;=\; q_{n-r-1,r}(x) + q_{n-r-1,r+1}(x)$$

and iterating, we get

$$q_{n-r,r}(x) \;=\; q_{n-r-1,r}(x) + q_{n-r-2,r+1}(x) + \cdots + q_{0,n-1}(x) + q_{0,n}(x) \qquad (3.13)$$

for $r \in \{0, 1, \ldots, n\}$. This means that the first $n+1$ terms of $\mathcal{Q}_n$ are the partial sums of the reverse of $\mathcal{Q}_{n-1}$ and hence they form an interlacing sequence, by part (b) of Lemma 1.57. Thus, by part (a) of this lemma, to complete the induction it suffices to show that $q_{n,0}(x)$ and $q_{0,n}(x)$ interlace $q_{0,n+1}(x)$. As already discussed, and in view of (3.11) and (3.12), this is equivalent to showing that $h_{\mathcal{A}}(\sigma_n, x) + x^n h_{\mathcal{A}}(\sigma_n, 1/x)$ and $h_{\mathcal{A}}(\sigma_n, x)$ interlace $h_{\mathcal{A}}(\sigma_{n+1}, x)$. To verify this we note that, setting $r = 0$ in Equation (3.13), comparing with (3.11) and (3.12) and replacing $n$ with $n + 1$, we get

$$q_{n,0}(x) + q_{n-1,1}(x) + \cdots + q_{0,n}(x) \;=\; h_{\mathcal{A}}(\sigma_{n+1}, x). \qquad (3.14)$$

Since the sum of the terms of an interlacing sequence is interlaced by the first term, we conclude that $h_{\mathcal{A}}(\sigma_n, x) + x^n h_{\mathcal{A}}(\sigma_n, 1/x)$ interlaces $h_{\mathcal{A}}(\sigma_{n+1}, x)$. Finally, applying part (c) of Lemma 1.57 to the interlacing sequence $\mathcal{Q}_{n-1}$ we conclude that the sum of the first $n$ terms of this sequence, which equals $h_{\mathcal{A}}(\sigma_n, x)$, interlaces the sum of the partial sums of the reverse of $\mathcal{Q}_{n-1}$, which equals $h_{\mathcal{A}}(\sigma_{n+1}, x)$. This completes the proof of the claim.

Finally, note that $x^n h_{\mathcal{A}}(\sigma_n, 1/x)$ and $x^{n+1} h_{\mathcal{A}}(\sigma_{n+1}, 1/x)$ are the last two terms of $\mathcal{Q}_n$. Since this sequence is interlacing, the two polynomials are real-rooted and the former interlaces the latter. As already discussed, this means that $h_{\mathcal{A}}(\sigma_n, x)$ is real-rooted and interlaces $h_{\mathcal{A}}(\sigma_{n+1}, x)$. Similarly, the sum of the first $n$ terms of the sequence $\mathcal{Q}_{n-1}$ interlaces the last term. In view of (3.12) and (3.14), this means that $h_{\mathcal{A}}(\sigma_n, x)$ interlaces $x^n h_{\mathcal{A}}(\sigma_n, 1/x)$ and, equivalently, that $h_{\mathcal{A}}(\sigma_n, x)$ is interlaced by $x^{n-1} h_{\mathcal{A}}(\sigma_n, 1/x)$. Since we already know from Proposition 3.7 that $h_{\mathcal{A}}(\sigma_n, x)$ has a nonnegative symmetric decomposition with respect to $n - 1$, this decomposition must be real-rooted and interlacing by [BS19, Theorem 2.6]. $\qquad\square$

| | $r = 0$ | $r = 1$ | $r = 2$ |
|---|---|---|---|
| $n = 0$ | $1$ | $x$ | $2x + x^2$ |
| $n = 1$ | $1 + x$ | $3x + x^2$ | $5x + 10x^2 + x^3$ |
| $n = 2$ | $1 + 4x + x^2$ | $8x + 11x^2 + x^3$ | $12x + 61x^2 + 30x^3 + x^4$ |
| $n = 3$ | $1 + 12x + 12x^2 + x^3$ | $20x + 72x^2 + 31x^3 + x^4$ | |

Table 3.1: Some polynomials $q_{n,r}(x)$.

Let us write $\theta_A(\sigma_n, x) := h_A(\sigma_n, x) - h_A(\partial\sigma_n, x)$. As mentioned in the proof of Proposition 3.7, the expression $h_A(\sigma_n, x) = h_A(\partial\sigma_n, x) + \theta_A(\sigma_n, x)$ is the symmetric decomposition of $h_A(\sigma_n, x)$ with respect to $n-1$. Thus, $h_A(\partial\sigma_n, x)$ and $\theta_A(\sigma_n, x)$ are real-rooted by Theorem 3.18. Although the latter appears to be a very special case of Conjecture 3.1, according to [Ath20b, Theorem 1.2], it would imply the conjecture if the following statement (which we have verified computationally for $n \leq 20$) also turns out to be true.

**Conjecture 3.19.** *The polynomial $h_A(\sigma_{n-1}, x)$ interlaces $\theta_A(\sigma_n, x)$ for every positive integer $n$.*

**Remark 3.20.** The polynomial $h_A(\sigma_n, x) + x^n h_A(\sigma_n, 1/x)$, shown to be real-rooted in the proof of Theorem 3.18, is equal to the $h$-polynomial of a flag triangulation of the $(n-1)$-dimensional sphere. Indeed, let $\Gamma = \mathrm{sd}_A(\sigma_n)$, so that $h(\Gamma, x) = h_A(\sigma_n, x)$. Then, in the notation of Section 3.1, in particular Remark 3.5, $\Delta = \Delta_A(\Gamma)$ is a flag triangulation of the $(n - 1)$-dimensional sphere and $h(\Delta, x) = h(\Gamma, x) + h^\circ(\Gamma, x) = h(\Gamma, x) + x^n h(\Gamma, 1/x) = h_A(\sigma_n, x) + x^n h_A(\sigma_n, 1/x)$.

**Remark 3.21.** The polynomial

$$\bar{p}_A(\sigma_n, x) := \sum_{k=0}^{n} \bar{p}_A(n, k)x^k = \sum_{k=0}^{n} c(n, k)x^k,$$

where $c(n, k)$ is the number of permutations in $\mathfrak{S}_n$ with excedance set equal to $[k]$, was shown to be symmetric and unimodal in [ES00, Section 3]. For the first few values of $n$,

$$\bar{p}_A(\sigma_n, x) = \begin{cases} 1, & \text{if } n = 1 \\ 1 + x, & \text{if } n = 2 \\ 1 + 3x + x^2, & \text{if } n = 3 \\ 1 + 7x + 7x^2 + x^3, & \text{if } n = 4 \\ 1 + 15x + 31x^2 + 15x^3 + x^4, & \text{if } n = 5 \\ 1 + 31x + 115x^2 + 115x^3 + 31x^4 + x^5, & \text{if } n = 6 \\ 1 + 63x + 391x^2 + 675x^3 + 391x^4 + 63x^5 + x^6, & \text{if } n = 7. \end{cases}$$

The following statement is stronger than the real-rootedness of $h_A(\sigma_n, x)$.

**Conjecture 3.22.** *The polynomial $\bar{p}_A(\sigma_n, x)$ is real-rooted and interlaces $\bar{p}_A(\sigma_{n+1}, x)$ for every $n \in \mathbb{N}$. In particular, $\bar{p}_A(\sigma_n, x)$ is $\gamma$-positive for every $n \in \mathbb{N}$.*

## 3.4 The local $h$-polynomial

We now focus on the local $h$-polynomial $\ell_A(\sigma_n, x)$ of the antiprism triangulation of $\sigma_n$. For the first few values of $n$,

$$
\ell_A(\sigma_n, x) = \begin{cases}
1, & \text{if } n = 0 \\
0, & \text{if } n = 1 \\
2x, & \text{if } n = 2 \\
3x + 3x^2, & \text{if } n = 3 \\
4x + 30x^2 + 4x^3, & \text{if } n = 4 \\
5x + 130x^2 + 130x^3 + 5x^4, & \text{if } n = 5 \\
6x + 435x^2 + 1460x^3 + 435x^4 + 6x^5, & \text{if } n = 6 \\
7x + 1281x^2 + 10535x^3 + 10535x^4 + 1281x^5 + 7x^6, & \text{if } n = 7.
\end{cases}
$$

We now provide a recurrence, combinatorial interpretations and formulas for the polynomials $\ell_A(\sigma_n, x)$.

**Proposition 3.23.** (a) *We have*

$$
\ell_A(\sigma_n, x) = \sum_{k=0}^{n-1} \binom{n}{k} \ell_A(\sigma_k, x) \left( (1+x)^{n-k} - 1 - x^{n-k} \right) \tag{3.15}
$$

*for every positive integer $n$. In particular, $\ell_A(\sigma_n, x)$ is unimodal for every $n \in \mathbb{N}$.*

(b) *The coefficient of $x^k$ in $\ell_A(\sigma_n, x)$ is equal to:*

- *the number of proper multi-pointed ordered set partitions of $[n]$ of weight $k$,*

- *the number of ways to choose an ordered set partition $\pi$ of $[n]$ and to color $k$ elements of $[n]$ black and the remaining $n - k$ white, so that no block of $\pi$ is monochromatic,*

- *the number of ordered set partitions $(B_1, B_2, \ldots, B_m)$ of $[n]$ having an even number of blocks for which the union $\bigcup_{i=1}^{m/2} B_i$ has exactly $k$ elements,*

- *$\binom{n}{k}$ times the number of derangements in $\mathfrak{S}_n$ with excedance set equal to $[k]$,*

- *the explicit expression*

$$
\binom{n}{k} \sum_{j \geq 1} (j!)^2 S(k, j) S(n - k, j),
$$

*where $S(n, k)$ are the Stirling numbers of the second kind.*

*Proof.* The recurrence of part (a) follows from Proposition 3.7, as a special case of Equation (3.3). The unimodality of $\ell_A(\sigma_n, x)$ follows directly from the recurrence by induction on $n$ (and, alternatively, from the regularity of the antiprism triangulation of the simplex; see the proof of 4.21).

For part (b), the first interpretation was already shown in the proof of Proposition 3.16 and the second is a restatement of the first. The third interpretation follows from the first and the proof of the corresponding result of Proposition 3.16 by noting that in the provided bijection the set $[n] \setminus S$ is always empty. Furthermore, the fifth interpretation follows from the second one since there are $\binom{n}{k}$ ways to choose the $k$ black elements of $[n]$ and for every such choice and every $j \geq 1$, there are $j! S(k, j) \cdot j! S(n - k, j)$ ways to choose an ordered partition of $[n]$ with $j$ blocks, none of which is monochromatic.

Finally, we deduce the fourth interpretation from the corresponding result of part (b) of Proposition 3.16. Let us use the notation adopted in the proof of that proposition, write $\ell_A(\sigma_n, x) = \sum_{k=0}^{n} \ell_A(n, k) x^k$ for $n \in \mathbb{N}$ and set

$$\ell_A(n, k) = \binom{n}{k} \bar{\ell}_A(n, k).$$

Then, by the second interpretation, considering the elements $1, 2, \ldots, k$ colored black and the other elements of $[n]$ colored white, $\bar{\ell}_A(n, k)$ is equal to the number of ordered set partitions of $[n]$ with no monochromatic block. By Proposition 3.16, under the same coloring convention, $\bar{p}_A(n, k)$ is equal to the number of ways to choose a set $[k] \subseteq S \subseteq [n]$ and an ordered set partition of $S$ with no monochromatic block. These interpretations imply that

$$\bar{p}_A(n, k) = \sum_{i=0}^{n-k} \binom{n-k}{i} \bar{\ell}_A(n - i, k)$$

for all $n, k$. Denoting by $d(n, k)$ the number of derangements in $\mathfrak{S}_n$ with excedance set equal to $[k]$, it should also be clear that

$$c(n, k) = \sum_{i=0}^{n-k} \binom{n-k}{i} d(n - i, k)$$

for all $n, k$. By Proposition 3.16, we have $\bar{p}_A(n, k) = c(n, k)$ for all $n, k$. Therefore, the two expressions for these numbers above and an easy induction show that $\bar{\ell}_A(n, k) = d(n, k)$ for all $n, k$ and the proof follows.

Alternatively, the fourth interpretation can be deduced by using so-called *P-decompositions* introduced in [KZ01].

For that, we need to introduce some more terminology. A *cycle* of length $k$ is a sequence $\sigma = s_1 \ldots s_k$ of $k$ distinct integers such that $s_1 = \min\{s_1, \ldots, s_k\}$. We call a cycle $\sigma$ *prime*, if there exists an index $i$, $2 \leq i \leq k$, such that $s_1 < s_2 < \cdots < s_{i-1} < s_i > s_{i+1} > \cdots > s_k$ and $s_{i-1} < s_k$. Hence a prime cycle is in particular unimodal. For a prime cycle $\sigma$, denote by $p(\sigma)$ the set of elements before the peak, i.e., $p(\sigma) = \{s_1, \ldots, s_{i-1}\}$ for $\sigma = s_1 \ldots s_k$ with peak at index $i$. As usual, to each cycle $\sigma = s_1 \ldots s_k$ we can associate a permutation $w$ on $\{s_1, \ldots, s_k\}$ by defining $w(s_i) = s_{i+1}$ for $i \in [k]$ where $s_{k+1} = s_1$.

For a composition $(l_1, \ldots, l_m)$ of $n$, we say that a *P-decomposition* of type $(l_1, \ldots, l_m)$ of $n$ is a sequence of prime cycles $\tau = (\tau_1, \ldots, \tau_m)$ such that $\tau_i$ is of length $l_i$ and the underlying sets of $\tau_i$ form a partition of $[n]$.

Using two algorithms (for details see [KZ01, Section 2]) one can associate to each derangement of $[n]$ a unique $P$-decomposition of $[n]$ [KZ01, Theorem 2.4]. We prove the claim by showing bijctions between the following sets:

(i) ordered set partitions of $[n]$ with no monochromatic block and exactly the elements $1, \ldots, k$ colored black,

(ii) $P$-decompositions $\tau = (\tau_1, \ldots, \tau_m)$ of $n$ such that $\bigcup_i p(\tau_i) = [k]$,

(iii) derangements $\pi$ of $[n]$ with $\mathrm{Exc}(\pi) = [k]$.

Let $(B_1, \ldots, B_m)$ be an ordered set partition described in (i). The elements in each block $B_i$ can be ordered uniquely in a unimodal way, which gives a cycle $\tau_i$ that is prime since all white elements are bigger than the black elements. Hence one obtains a $P$-decomposition as described in (ii). This gives the first bijection.

Let $\pi$ be a derangement with excedance set $\mathrm{Exc}(\pi) = [k]$. By exploiting the effect of the $U$-algorithm defined in [KZ01, Section 2] on these derangements we prove the second bijection.

Let $C(\pi) = (\tau_1, \ldots, \tau_m)$ be the derangement $\pi$ factorized in cycles $\tau_i$. Since $\pi$ is a derangement each of them has length at least two. For each cycle $\tau_i = s_{i_1} \cdots s_{i_l}$ it holds that

$$s_{i_j} < s_{i_{j+1}} \Leftrightarrow s_{i_j} \in [k]. \tag{3.16}$$

The $U$-algorithm decomposes a cycle $\tau = s_1 \cdots s_l$ in unimodal cycles in the following way:

1. If the cycle $\tau$ is unimodal, the cycle does not get changed.

2. Otherwise, there exists a largest index $i$ such that $s_{i-1} > s_i < s_{i+1}$. Let $j$ be the unique index greater than $i$ such that $s_j > s_i > s_{j+1}$, where $s_{l+1} = s_1$. Then $s_i \ldots s_j$ is a unimodal cycle, and we decompose $\tau$ into $U(\tau) = (U(s_1 \cdots s_{i-1} s_{j+1} \cdots s_l), s_i \cdots s_j)$, so we get a sequence of unimodal cycles.

Applying this algorithm to each cycle $\tau_i$ of $C(\pi)$, we can associate to each derangement a sequence of unimodal cycles $(u_1, \ldots, u_m)$. From (3.16) we conclude that $\bigcup_i p(u_i) = [k]$. Furthermore, each of the cycles $u_i$ is already a prime cycle, since all the elements after the peak have to be greater than $k$. So we get a $P$-decomposition of $[n]$ as described above. $\qquad\square$

Following the notation introduced in the previous proof, we set $\bar{\ell}_A(\sigma_n, x) := \sum_{k=0}^{n} \ell_A(n, k) x^k$. We note that, since the polynomial $\ell_A(\sigma_n, x)$ is symmetric with center of symmetry $n/2$, so is $\bar{\ell}_A(\sigma_n, x)$.

**Conjecture 3.24.**   (a) *The polynomial $\ell_A(\sigma_n, x)$ is real-rooted and it interlaces $\ell_A(\sigma_{n+1}, x)$ for every $n \in \mathbb{N}$. In particular, $\ell_A(\sigma_n, x)$ is $\gamma$-positive for every $n \in \mathbb{N}$.*

(b) *The polynomial $\bar{\ell}_A(\sigma_n, x)$ is real-rooted and interlaces $\bar{\ell}_A(\sigma_{n+1}, x)$ for every $n \in \mathbb{N}$. In particular, $\bar{\ell}_A(\sigma_n, x)$ is $\gamma$-positive for every $n \in \mathbb{N}$.*

A combinatorial interpretation of the polynomial $\theta_A(\sigma_n, x) := h_A(\sigma_n, x) - h_A(\partial \sigma_n, x)$, which appeared in Conjecture 3.19, can be deduced from Proposition 3.23.

**Corollary 3.25.** *The coefficient of $x^k$ in $\theta_A(\sigma_n, x)$ equals the number of ways to choose an ordered set partition $\pi$ of $[n]$ and to color $k$ elements of $[n]$ black and the remaining $n - k$ white, so that no block of $\pi$ is monochromatic and there is a black element which is larger than a white element in the last block of $\pi$.*

*Proof.* As a consequence of [JKMS19, Lemma 4.1] (and as already discussed in the proof of Proposition 3.7), we have

$$\theta_A(\sigma_n, x) \;=\; \ell_A(\sigma_n, x) \;-\; \sum_{m=0}^{n-2} \binom{n}{m} \ell_A(\sigma_m, x)(x + x^2 + \cdots + x^{n-m-1})$$

for every positive integer $n$. By the second interpretation of $\ell_A(\sigma_n, x)$ provided by Proposition 3.23, the coefficient of $x^k$ in the sum on the right-hand side is equal to the number of ways to choose an ordered set partition $\pi$ of $[n]$ and to color $k$ elements of $[n]$ black and the remaining $n - k$ white, so that no block of $\pi$ is monochromatic and every black element in the last block of $\pi$ is smaller than every white element of that block. Thus, the proposed interpretation of $\theta_A(\sigma_n, x)$ follows from the previous equation and Proposition 3.23. $\qquad\square$

## 3.5 Face-vector transformations

The general results of [Ath20b] on uniform triangulations of simplicial complexes imply that there exist nonnegative integers $q_A(n, k)$ and $p_A(n, k, j)$ for $n, k, j \in \mathbb{N}$ with $k, j \leq n$ such that

$$f_{j-1}(\mathrm{sd}_A(\Delta)) \;=\; \sum_{k=j}^{n} q_A(k, j) f_{k-1}(\Delta) \tag{3.17}$$

and

$$h_j(\mathrm{sd}_A(\Delta)) \;=\; \sum_{k=0}^{n} p_A(n, k, j) h_k(\Delta) \tag{3.18}$$

for every $(n - 1)$-dimensional simplicial complex $\Delta$ and every $j \in \{0, 1, \ldots, n\}$. The former equation is easy to explain; a simple counting argument (see the proof of [Ath20b, Theorem 4.1]) shows its validity when $q_A(n, k)$ is defined as the number of $(k - 1)$-dimensional faces in the interior of the antiprism triangulation of $\sigma_n$. This yields the following statement.

**Proposition 3.26.** *For all integers $n \geq 1$ and $k \in \{0, 1, \ldots, n\}$, $q_A(n, k)$ is equal to the number of multi-pointed ordered set partitions of $[n]$ of weight $k$. Moreover, we have the explicit formula*

$$q_A(n, k) = \binom{n}{k} \sum_{j=0}^{k} j! S(k, j) j^{n-k}$$

*where, as usual, $S(k, j)$ is a Stirling number of the second kind.*

*Proof.* The proposed combinatorial interpretation follows from our previous discussion and that in Section 3.2. To verify the formula, we note that there are $\binom{n}{k} \cdot j! S(k, j)$ ways to chose a $k$-element subset $S$ of $[n]$ and an ordered partition of $S$ with $j$ blocks and, for each such choice there are $j^{n-k}$ ways to distribute the remaining $n - k$ elements of $[n]$ in the blocks so as to form a multi-pointed ordered set partition of $[n]$ with set of chosen elements equal to $S$. $\square$

Equation (3.18) and the nonnegativity of the coefficients $p_A(n, k, j)$ which appear there are less obvious. As described in Section 2.3, an explicit formula and a recurrence are given for these numbers in [Ath20b] in the general framework of uniform triangulations.

In particular we have that the recurrence (2.6)

$$p_A(n, k, j) = p_A(n, k - 1, j) + p_A(n - 1, k - 1, j - 1) - p_A(n - 1, k - 1, j) \quad (3.19)$$

holds for all $k, j \in \{0, 1, \ldots, n\}$ with $k \geq 1$. We will also prove this recurrence directly from the explicit formula for $p_A(n, k, j)$ obtained in Proposition 3.27.

We will keep in mind that $p_A(n, 0, j) = p_A(n, j)$ is the coefficient of $x^j$ in $h_A(\sigma_n, x)$. This observation is the special case $\Delta = \sigma_n$ of Equation (3.18).

The following combinatorial interpretations of $p_A(n, k, j)$ generalize some of those given for $p_A(n, k)$ in Proposition 3.16.

**Proposition 3.27.** *For all integers $n \geq 1$ and $k \in \{0, 1, \ldots, n\}$, $p_A(n, k, j)$ we have the explicit expression*

$$p_A(n, k, j) = \sum_{l=0}^{n} \sum_{i=0}^{j} (-1)^{j-i} \binom{n-i}{n-j} \binom{n-k}{n-l} q_A(l, i). \quad (3.20)$$

*Moreover, $p_A(n, k, j)$ is equal to:*

- *the number of ways to choose a set $[k] \subseteq S \subseteq [n]$ and an ordered set partition $\pi$ of $S$ and to color $j$ elements of $S$ black and the remaining elements white, so that the following condition holds: if a block $B$ of $\pi$ is monochromatic, then*

  - *$B$ is the first block of $\pi$,*

  - *$B \subseteq [k]$, and*

  - *all elements of $B$ are colored black.*

- *the number of ordered set partitions $(B_1, B_2, \ldots, B_m)$ of $[n]$ for which the following conditions hold:*
  - *if $m$ is even, then $\bigcup_{i=1}^{\lfloor m/2 \rfloor} B_i$ has exactly $j$ elements, and*
  - *if $m$ is odd, then the union of $\bigcup_{i=1}^{\lfloor m/2 \rfloor} B_i$ and $B_m \cap [k]$ has exactly $j$ elements.*

*Proof.* We first proof the explicit expression of the numbers $p_A(n, k, j)$ by using the transformations of the $h$- and $f$-vectors into each other in Proposition 1.13.

$$
\begin{aligned}
h_j(\mathrm{sd}_A(\Delta)) &= \sum_{i=0}^{j} (-1)^{j-i} \binom{n-i}{n-j} f_{i-1}(\mathrm{sd}_A(\Delta)) \\
&= \sum_{i=0}^{j} (-1)^{j-i} \binom{n-i}{n-j} \sum_{l=i-1}^{n-1} f_l(\Delta) q_A(l+1, i) \\
&= \sum_{i=0}^{j} \sum_{l=i}^{n} (-1)^{j-i} \binom{n-i}{n-j} q_A(l, i) f_{l-1}(\Delta) \\
&= \sum_{i=0}^{j} \sum_{l=i}^{n} (-1)^{j-i} \binom{n-i}{n-j} q_A(l, i) \sum_{k=0}^{l} \binom{n-k}{n-l} h_k(\Delta) \\
&= \sum_{k=0}^{n} \left[ \sum_{l=0}^{n} \sum_{i=0}^{j} (-1)^{j-i} \binom{n-i}{n-j} \binom{n-k}{n-l} q_A(l, i) \right] h_k(\Delta) \\
&= \sum_{k=0}^{n} p_A(n, k, j) h_k(\Delta).
\end{aligned}
$$

Now we prove the first combinatorial interpretation. Let $Q(n, k, j)$ be the collection of triples of sets $S$, partitions of $S$ and colorings of the elements of $S$ described in the first proposed combinatorial interpretation of $p_A(n, k, j)$ and let $q(n, k, j)$ be the cardinality of $Q(n, k, j)$. We will show that $p_A(n, k, j) = q(n, k, j)$. This is true for $k = 0$ by the first combinatorial interpretation of $p_A(n, j) = p_A(n, 0, j)$ provided by Proposition 3.16. Thus, it suffices to show that the numbers $q(n, k, j)$ satisfy recurrence (3.19) or, equivalently, that

$$
q(n, k - 1, j) \;=\; q(n, k, j) + q(n - 1, k - 1, j) - q(n - 1, k - 1, j - 1)
$$

for $k \geq 1$. By definition, $q(n, k - 1, j)$ is the number of triples in the collection $Q(n, k - 1, j)$, each one consisting of a set $S$, a partition of $S$ and a coloring of the elements of $S$ having certain properties. Clearly, we have $k \notin S$ for exactly $q(n - 1, k - 1, j)$ of these triples. Moreover, we have $k \in S$ for exactly $q(n, k, j) - q(n - 1, k - 1, j - 1)$ of them, since for exactly $q(n - 1, k - 1, j - 1)$ of the triples in $Q(n, k, j)$ there is a monochromatic block which contains $k$. This proves the first interpretation.

As an alternative proof, by computing the coefficient of $x^j$ in the right-hand side of [Ath20b, Equation (12)] we get the explicit expression

$$p_A(n,k,j) = \sum_{r=0}^{n} \sum_{i=0}^{j} \binom{k}{i} \binom{n-k}{n-r-i} \ell_A(r, j-i).$$

We will show that the double sum on the right side is also equal to $q(n, k, j)$. We count the number of ways to choose choose a set $S$, a partition $\pi$ and coloring as in the statement of the proposition. If the monochromatic block $B$ has exactly $i$ elements, when present, and there is a total of $r$ elements in the remaining blocks of $\pi$, there are $\binom{k}{i}\binom{n-k}{n-r-i}$ ways to choose the $i$ elements of $B$ and the $n-r-i$ elements of $[n]$ not in the blocks of $\pi$. For each such choice, by Proposition 3.23, there are $\ell_A(r, j-i)$ ways to choose the blocks of $\pi$ other than $B$.

Alternatively, we can prove this interpretation using an involution, see for example [Sta12, Section 2.6] for this concept.
By the explicit expression above it holds

$$\begin{aligned}
p_A(n,k,j) &= \sum_{l=k}^{n} \binom{n}{l} \sum_{i=0}^{j} (-1)^{j-i} \binom{n-i}{n-j} q_A(l,i) \qquad\qquad (3.21)\\
&= \sum_{l=k}^{n} \binom{n-k}{l-k} q_A(l,j) - \sum_{l=k}^{n} (n-j+1) \binom{n-k}{l-k} q_A(l,j-1)\\
&\quad + \sum_{l=k}^{n} \binom{n-j+2}{2} \binom{n-k}{l-k} q_A(l,j-2) - \ldots\\
&= \sum_{l=0}^{n-k} \binom{n-k}{l} q_A(l+k,j) - \sum_{l=0}^{n-k} \binom{n-k}{l} (n-j+1) q_A(l+k,j-1)\\
&\quad + \sum_{l=0}^{n-k} \binom{n-k}{l} \binom{n-j+2}{2} q_A(l+k,j-2) - \ldots
\end{aligned}$$

For fixed nonnegative integers $n, j$ and $k$, take all tuples of ordered partitions $\sigma = (B_1, \ldots, B_r)$ of any $[k] \subseteq S \subseteq [n]$ and colorings $c : [n] \to \{0,1,2\}$, such that $c([n] \setminus S) \subseteq \{1,2\}$, $|c^{-1}(\{0,1\})| = j$, and for all $1 \le l \le r$ there is at least one element $x \in B_l$ with $c(x) = 0$. Denote the set of all these tuples $X_{n,j}^k$. Remark that the elements, in which the map $c$ is such that its image is only contained in $\{0,1\}$, correspond to the elements in $Q(n,k,j)$.

For $\omega = (c, \sigma) \in X_{n,j}^r$ with $\sigma = (B_1, \ldots, B_r)$ let $m(\omega)$ be the maximal $i \in [n]$ satisfying one of the following conditions:

1. $c(i) = 1$,

2. $c(i) = 0$, $i \in B_l$ and there exist an $m \in B_l$ with $c(m) = 1$,

3. $c(i) = 0$, $i \in B_l$ with $l \ne 1$, and $c(m) = 0$ for all $m \in B_l$,

   4. $c(i) = 0$, $i \in B_1$, $c(m) = 0$ for all $m \in B_1$ and $i > k$.

If there is no $i \in [n]$ satisfying one of these conditions, define $m(\omega) = 0$. We define the map $\tau : X_{n,j}^k \to X_{n,j}^k$ as follows:

- If $m(\omega)$ satisfies condition 1 and $m(\omega) \notin B_l$ for every $l$, then $\tau(\omega) = (c', \sigma')$ with $\sigma' = (\{m(\omega)\}, B_1, \ldots, B_r)$ and $c'(m(\omega)) = 0$ and $c'(i) = c(i)$ for $i \in [n] \setminus m(\omega)$.

- If $m(\omega)$ satisfies condition 1 and $m(\omega) \in B_l$ for some $l$, then $\tau(\omega) = (c', \sigma')$ with $c'(m(\omega)) = 0$ and $\sigma' = (B1, \ldots, B_{l-1}, B_l \setminus m(\omega), \{m(\omega)\}, B_{l+1}, \ldots, B_r)$.

- If $m(\omega)$ satisfies condition 2, or $m(\omega)$ satisfies condition 3 or 4 and is not in a singleton block, let $\tilde{m}(\omega)$ be the minimal $i \in B_l$ with $c(i) \neq 2$. Then $\tau(\omega) = (c', \sigma)$ with $c'(\tilde{m}(\omega)) = 1 - c(\tilde{m}(\omega))$ and $c'(i) = c(i)$ for $i \in [n] \setminus \tilde{m}(\omega)$.

- If $m(\omega)$ satisfies condition 3 and $B_l$ is a singleton block, then $\tau(\omega) = (c', \sigma')$ with $\sigma' = (B_1, \ldots, B_{l-1} \cup m(\omega), B_{l+1}, \ldots, B_r)$ and $c'(m(\omega)) = 1$ and $c'(i) = c(i)$ for $i \in [n] \setminus m(\omega)$.

- If $m(\omega)$ satisfies condition 4 and $B_1$ is a singleton block, then $\tau(\omega) = (c', \sigma')$ with $\sigma' = (B_2, \ldots, B_k)$ and $c'(m(\omega)) = 1$ and $c'(i) = c(i)$ for $i \in [n] \setminus m(\omega)$.

- If $m(\omega) = 0$, then $\tau(\omega) = \omega$.

The map $\tau$ is an involution on $X_{n,j}^k$, such that if $\tau(\omega) = (c', \sigma')$, then $|c'^{-1}(1)| = |c^{-1}(1)| \pm 1$, and if $\tau(\omega) = \omega$, then $|c^{-1}(1)| = 0$.

The sum in Equation (3.21) can be interpreted as

$$p_A(n, k, j) = \sum_{i=0}^{j} (-1)^i |\{\omega \in X_{n,j}^k : |c^{-1}(1)| = i\}|,$$

which shows, that $p_A(n, k, j)$ is the number of fixpoints of the involution $\tau$, which is

$$p_A(n, k, j) = |Q(n, k, j)| = q(n, k, j).$$

To prove the second interpretation, it suffices to find a bijection from $Q(n, k, j)$ to the collection of ordered set partitions described there. Such a bijection can be constructed as an obvious extension of the one provided in the proof of Proposition 3.16 for the special case $k = 0$. More specifically, the elements of the monochromatic block, if present, of a colored ordered partition in $Q(n, k, j)$ should be included in the last block of the ordered partition produced by the bijection; the details are left to the interested reader. $\qquad\square$

As mentioned before, we can prove the recursion in (2.6) only using the explicit expression (3.20) in the following way: Let $P := p_A(n, k-1, j) + p_A(n-1, k-1, j-$

$1) - p_{\mathcal{A}}(n-1, k-1, j)$. Then

$$
\begin{aligned}
P &= \sum_{l=0}^{n}\sum_{i=0}^{j}(-1)^{j-i}\binom{n-i}{n-j}\binom{n-k+1}{n-l}q_{\mathcal{A}}(l,i) \\
&\quad + \sum_{l=0}^{n-1}\sum_{i=0}^{j-1}(-1)^{j-1-i}\binom{n-1-i}{n-j}\binom{n-k}{n-1-l}q_{\mathcal{A}}(l,i) \\
&\quad - \sum_{l=0}^{n-1}\sum_{i=0}^{j}(-1)^{j-i}\binom{n-1-i}{n-1-j}\binom{n-k}{n-1-l}q_{\mathcal{A}}(l,i) \\
&= \sum_{l=0}^{n-1}\sum_{i=0}^{j-1}(-1)^{j-i}\left[\binom{n-i}{n-j}\binom{n-k+1}{n-l} - \right. \\
&\qquad\qquad \left. \left(\binom{n-1-i}{n-j}+\binom{n-1-i}{n-1-j}\right)\binom{n-k}{n-1-l}\right]q_{\mathcal{A}}(l,i) \\
&\quad + \sum_{i=0}^{j}(-1)^{j-i}\binom{n-i}{n-j}q_{\mathcal{A}}(n,i) + \sum_{l=0}^{n-1}\left[\binom{n-k+1}{n-l}-\binom{n-k}{n-1-l}\right]q_{\mathcal{A}}(l,j) \\
&= \sum_{l=0}^{n-1}\sum_{i=0}^{j-1}(-1)^{j-i}\binom{n-i}{n-j}q_{\mathcal{A}}(l,i) + \sum_{i=0}^{j}(-1)^{j-i}\binom{n-i}{n-j}q_{\mathcal{A}}(n,i) + \sum_{l=0}^{n-1}\binom{n-k}{n-l}q_{\mathcal{A}}(l,j) \\
&= \sum_{l=0}^{n}\sum_{i=0}^{j}(-1)^{j-i}\binom{n-i}{n-j}\binom{n-k}{n-l}q_{\mathcal{A}}(l,i) \\
&= p_{\mathcal{A}}(n,k,j).
\end{aligned}
$$

# 4 The Lefschetz property for the antiprism triangulation

The theory of Lefschetz properties is a topic arising in various areas of mathematics, such as algebraic topology, algebraic geometry, commutative algebra, representation theory and combinatorics. In [MN13] and [HMM$^+$13], one can find an overview of the wide variety of the fields in which Lefschetz properties are studied and found. Lefschetz properties are also an important tool in the context of face enumeration of simplicial complexes. They were used by Stanley to prove the classical $g$-theorem, providing a characterization of the $g$-vectors, an invariant directly related to the $h$-vector, of boundaries of simplicial polytopes. Since then, also for several other classes of simplicial complexes Lefschetz properties have been proven, some of which we will review in the following. Having a Lefschetz property for a class of simplicial complexes yields certain enumerative consequences for the $h$-vector of these complexes. One example of this in the context of subdivisions is again the barycentric subdivision, in [KN09] it was shown that the barycentric subdivision of a shellable complex has the almost strong Lefschetz property, showing that its $h$-vector is unimodal with the peak in the middle.

We also prove the unimodality of $h(\mathrm{sd}_A(\Delta))$ for every Cohen–Macaulay simplicial complex $\Delta$ and show that the peak appears in the middle, by studying Lefschetz properties of the Stanley–Reisner ring of $\mathrm{sd}_A(\Delta)$. The following result is an analogue of the main result of [KN09, Theorem 1.1] for the barycentric subdivision.

**Theorem 4.1.** *The complex* $\mathrm{sd}_A(\Delta)$ *has the almost strong Lefschetz property over* $\mathbb{R}$ *for every shellable simplicial complex* $\Delta$.

*Moreover, for every* $(n-1)$-*dimensional Cohen–Macaulay simplicial complex* $\Delta$, *the $h$-vector of* $\mathrm{sd}_A(\Delta)$ *is unimodal, with the peak being at position $n/2$, if $n$ is even, and at $(n-1)/2$ or $(n+1)/2$, if $n$ is odd.*

## 4.1 The Lefschetz property: Definitions and general results

This section reviews basic definitions and background on Lefschetz properties for simplicial complexes and includes some preliminary technical results, which will be applied in the following section in the context of antiprism triangulations.

In the first part of this section we state the famous $g$-theorem for simplicial polytopes by Billera, Lee and Stanley, which was the starting point of the study of Lefschetz properties in this context.

The second part contains the algebraic definitions of the Lefschetz property. We review the implications on the $h$-vector of simplicial complexes having the Lefschetz property, and also state some known examples of classes of simplicial complexes with the Lefschetz property, as well as constructions under which the Lefschetz property is preserved.

### 4.1.1 The $g$-theorem

In the previous Chapters 2 and 3 we studied the face numbers of simplicial complexes, e.g., the nonnegativity of the $h$-vector. Another useful invariant for face enumeration which is directly related to the $h$-vector of a simplicial complex is the $g$-vector.

**Definition 4.2.** Let $\Delta$ be an $(n-1)$-dimensional simplicial complex and $h(\Delta) = (h_0(\Delta), \ldots, h_n(\Delta))$ its $h$-vector. Then the vector $g(\Delta) = (g_0(\Delta), \ldots, g_{\lfloor \frac{n}{2} \rfloor}(\Delta))$ with $g_0(\Delta) = 1$ and $g_i(\Delta) = h_i(\Delta) - h_{i-1}(\Delta)$, for $1 \leq i \leq \lfloor \frac{n}{2} \rfloor$, is called the $g$-vector of $\Delta$.

Hence for simplicial complexes having a symmetric $h$-vector, e.g., simplicial spheres, one can recover the $h$-vector and therefore the $f$-vector from its $g$-vector.

This is the reason that, instead of asking for a characterization of the $f$- or $h$-vector, one can equivalently ask for characterizations of the $g$-vector of complexes. A long standing conjecture regarding this characterization was raised by McMullen [McM71], for which we need the notion of an $M$-sequence.

Given the $i$-binomial representation of an integer $a$ (see Lemma 1.35), and similarly to the earlier notation $a^{(i)}$ in Section 1.1.4, we define $a^{\langle i \rangle}$ to be

$$a^{\langle i \rangle} = \binom{m_i + 1}{i + 1} + \binom{m_{i-1} + 1}{i} + \cdots + \binom{m_l + 1}{l + 1}.$$

Macaulay gave the following numerical characterization of $M$-sequences.

**Theorem 4.3.** [BH93, Theorem 4.2.10] *Let $a = (a_0, \ldots, a_s)$ be a sequence of non-negative integers. Then the following conditions are equivalent:*

*(i) $a$ is an $M$-sequence,*

*(ii) $a_0 = 1$ and $a_{i+1} \leq a_i^{\langle i \rangle}$ for $0 \leq i \leq s - 1$.*

**Remark 4.4.** Let $a = (a_0, \ldots, a_s)$ be a sequence of non-negative integers. Then the sequence is an $M$-sequence if and only if $a$ is the Hilbert function of a standard graded $\mathbb{F}$-algebra with a field $\mathbb{F}$.

McMullen [McM71] raised the following conjecture on the characterization of the $g$-vector of simplicial spheres (originally for boundaries of simplicial polytopes, and later extended to simplicial spheres).

**Conjecture 4.5.** [McM71] *Let $\Delta$ be a simplicial sphere. Then its $g$-vector is an $M$-sequence.*

An affirmative answer to this question was first given in the case, where $\Delta$ is the boundary of a simplicial polytope, this result is called the $g$-theorem (for simplicial polytopes). Billera and Lee [BL81] were able to construct a simplicial polytope, such that its boundary has a given $M$-sequence as $g$-vector. That the $g$-vector of the boundary of any simplicial polytope is an $M$-sequence was proven by Stanley [Sta80], using the hard Lefschetz theorem for toric varieties.

**Theorem 4.6.** [BL81, Sta80]

Let $h = (h_0, \dots, h_n) \in \mathbb{N}^{n+1}$ and $g = (1, h_1 - h_0, \dots, h_{\lfloor \frac{n}{2} \rfloor} - h_{\lfloor \frac{n}{2} \rfloor - 1})$. Then $h$ is the $h$-vector of the boundary complex of a simplicial polytope if and only if $g$ is an $M$-sequence.

Recently, Adiprasito [Adi18] announced a proof of the $g$-conjecture in full generality for simplicial spheres.

### 4.1.2 Lefschetz properties

In this part, we state the definitions of the Lefschetz property, and review some constructions, under which Lefschetz properties are preserved for simplicial complexes. Furthermore we introduce the contraction of an edge and the strong Link condition, which is needed for the proofs in the following section.

**Definition 4.7.** Let $\Delta$ be an $(n-1)$-dimensional simplicial complex. Let $\Theta = \theta_1, \dots, \theta_n$ be linear forms of degree $1$ in $\mathbb{F}[\Delta]$. Then $\Theta$ is a *linear system of parameters* (l.s.o.p. for short) for $\mathbb{F}[\Delta]$ if the quotient $\mathbb{F}[\Delta]/\Theta\mathbb{F}[\Delta]$ has finite dimension as a vector space over $\mathbb{F}$.

Assuming that $\mathbb{F}$ is an infinite field, the existence of a l.s.o.p. for $\mathbb{F}[\Delta]$ follows by the Noether normalization lemma [BH93, Theorem 1.5.17].

Having this definition at hand, we can also give an alternative definition for the Cohen–Macaulay property defined in Definition 1.27, by saying a complex is Cohen–Macaulay over $\mathbb{F}$ if $\mathbb{F}[\Delta]$ is a free module over the polynomial ring $\mathbb{F}[\theta_1, \dots, \theta_n]$ for some l.s.o.p. $\theta_1, \dots, \theta_n$ for $\mathbb{F}[\Delta]$.

We now consider the algebraic side of the $g$-theorem, used in the proof of Stanley. For this we need some more definitions.

**Definition 4.8.** Let $\Delta$ be an $(n-1)$-dimensional simplicial complex which is Cohen–Macaulay over an infinite field $\mathbb{F}$ and let $s \leq n$ be a positive integer. We say that $\Delta$ has the $s$-*Lefschetz property* (over $\mathbb{F}$) if there exists a linear system of parameters $\Theta$ for $\mathbb{F}[\Delta]$ and a linear form $\omega \in \mathbb{F}[\Delta]$, such that the multiplication maps

$$\cdot\omega^{s-2i} : (\mathbb{F}[\Delta]/\Theta\mathbb{F}[\Delta])_i \to (\mathbb{F}[\Delta]/\Theta\mathbb{F}[\Delta])_{s-i}$$

are injective for all $0 \leq i \leq \lfloor (s-1)/2 \rfloor$.

Following [KN09], we call $\Delta$ *almost strong Lefschetz* (over $\mathbb{F}$) if it has the $(n-1)$-Lefschetz property. Usually, if $\Delta$ has the $n$-Lefschetz property and, additionally, the above multiplication maps are isomorphisms, one says that $\Delta$ is *strong Lefschetz* (or $\Delta$ has the *strong Lefschetz property*).

There is the following result on Lefschetz elements, it can be found in [Swa06, Proposition 3.6].

**Theorem 4.9.** *Let $\Delta$ be an $(n-1)$-dimensional Cohen–Macaulay complex on vertex set $[m]$. Let $G(\Delta)$ be the set of all pairs $(\Theta, \omega) \subseteq \mathbb{F}^{(n+1)m}$ such that $\Theta$ is an l.s.o.p. for $\mathbb{F}[\Delta]$ and $\omega$ is an n-Lefschetz element for $\mathbb{F}[\Delta]/\Theta\mathbb{F}[\Delta]$.*
*Then $G(\Delta)$ is a Zariski-open set.*

Following the proof of this result, one can show the analogous result for general $s$-Lefschetz elements.

Hence, if $\Delta$ has a Lefschetz property, then a generic l.s.o.p. and a generic linear form have this property.

In these algebraic terms, Stanley's part of the proof of the $g$-theorem yields the following result.

**Theorem 4.10.** *Let $\Delta$ be the boundary complex of a simplicial polytope and let $\mathbb{F}$ be a field of characteristic 0. Then $\Delta$ is strong Lefschetz over $\mathbb{F}$.*

As already mentioned, the Lefschetz property can be used to deduce numerical information on the $h$-vector. In order to connect the Lefschetz property with the $h$-vector of a simplicial complex, we require a result proved by Stanley.

**Theorem 4.11.** *Let $\Delta$ be an $(n-1)$-dimensional homology sphere or homology ball, and let $\Theta = \{\theta_1, \ldots, \theta_n\}$ be an l.s.o.p. for $\mathbb{F}[\Delta]$. Then*

$$\mathrm{Hilb}(\mathbb{F}[\Delta]/\Theta\mathbb{F}[\Delta], x) = \sum_{i=0}^{n} h_i(\Delta)x^i.$$

This means, we have

$$\dim(\mathbb{F}[\Delta]/\Theta\mathbb{F}[\Delta])_i = h_i(\Delta)$$

for all $i$. This shows in particular, that the $h$-vector of a homology sphere or ball is nonnegative. Together with a Lefschetz property one can infer even more information.

Suppose $\Delta$ has the strong Lefschetz property. Then for $0 \leq i \leq \lfloor \frac{n-1}{2} \rfloor$ the multiplication $\cdot\omega^{n-2i} : (\mathbb{F}[\Delta]/\Theta\mathbb{F}[\Delta])_i \to (\mathbb{F}[\Delta]/\Theta\mathbb{F}[\Delta])_{n-i}$ is an isomorphism, and hence the multiplication $\cdot\omega : (\mathbb{F}[\Delta]/\Theta\mathbb{F}[\Delta])_{i-1} \to (\mathbb{F}[\Delta]/\Theta\mathbb{F}[\Delta])_i$ is an injection. Together with Theorem 4.11, we get $h_{i-1}(\Delta) \leq h_i(\Delta)$.

Having this result in mind, Lefschetz properties are an important tool in the area of face enumeration of simplicial complexes; various classes of simplicial complexes are known to have such properties, here we just give some examples.

One way towards the $g$-conjecture, was to study if Lefschetz properties are preserved under a certain construction changing the simplicial complex. In [BN10], Babson and Nevo observed the effect on the Lefschetz property under several constructions on simplicial complexes, and proved the following.

**Proposition 4.12.** [BN10, Theorem 1.2] *Let $\Delta, \Gamma$ be homology spheres over an infinite field $\mathbb{F}$ and let $F \in \Delta$ be a face.*

*(i)* *If $\Delta$ and $\Gamma$ are strong Lefschetz over $\mathbb{F}$ and the characteristic of $\mathbb{F}$ is zero, then $\Delta * \Gamma$ is a strong Lefschetz homology sphere.*

*(ii)* *If $\Delta$ and $\mathrm{lk}_\Delta(F)$ are strong Lefschetz over $\mathbb{F}$ and the characteristic of $\mathbb{F}$ is zero, then $\mathrm{stellar}(\Delta, F)$ is a strong Lefschetz homology sphere.*

The following result is a special case of the main result of [KN09] which will be used in Section 4.2.

**Proposition 4.13.** [KN09, Proposition 2.3] *Let $\mathbb{F}$ be an infinite field. The barycentric subdivision of the simplex $\sigma_n$ is almost strong Lefschetz over $\mathbb{F}$.*

As seen in Proposition 4.12, Lefschetz properties are known to be preserved under certain constructions. Moreover, they behave well under sufficiently nice edge subdivisions. Before we can make this more precise, we need to introduce some definitions.

**Definition 4.14.** Let $\Delta$ be a simplicial complex on a vertex set $V$ which is endowed with a total order $<$. Given an edge $e = \{a, b\} \in \Delta$ with $a < b$, the *contraction* $\mathcal{C}_\Delta(e)$ of $\Delta$ with respect to $e$ is the simplicial complex on the vertex set $V \smallsetminus \{b\}$ which is obtained from $\Delta$ by identifying vertices $a$ and $b$, i.e.,

$$\mathcal{C}_\Delta(e) := \{F \in \Delta \ : \ b \notin F\} \cup \{(F \smallsetminus \{b\}) \cup \{a\} \ : \ b \in F \in \Delta\}.$$

**Definition 4.15.** We say that $\Delta$ satisfies the *Link Condition* with respect to $e$ if

$$\mathrm{lk}_\Delta(e) \ = \ \mathrm{lk}_\Delta(\{a\}) \cap \mathrm{lk}_\Delta(\{b\}).$$

We remark, that the inclusion $\mathrm{lk}_\Delta(e) \subset \mathrm{lk}_\Delta(\{a\}) \cap \mathrm{lk}_\Delta(\{b\})$ always holds, since for a face $F \in \Delta$ with $F \in \mathrm{lk}_\Delta(e)$, it holds $F \cup \{a, b\} \in \Delta$ and $F \cap \{a, b\} = \emptyset$, hence also $F \cup \{a\} \in \Delta$ and $F \cap \{a\} = \emptyset$, i.e., $F \in \mathrm{lk}_\Delta(\{a\})$, and for vertex $b$ respectively. For the boundary of the 2-simplex, the Link Condition does not hold for any of its three edges, e.g., $\mathrm{lk}_{\partial \sigma_3}(\{1, 2\}) = \{\emptyset\}$, while $\mathrm{lk}_{\partial \sigma_3}(\{1\}) \cap \mathrm{lk}_{\partial \sigma_3}(\{2\}) = \{\{3\}, \emptyset\}$.

This means, that the crucial part for checking the Link Condition is the inclusion $\mathrm{lk}_\Delta(e) \supset \mathrm{lk}_\Delta(\{a\}) \cap \mathrm{lk}_\Delta(\{b\})$.

**Proposition 4.16.** *Let $\mathbb{F}$ be an infinite field and let $\Delta$ be an $(n-1)$-dimensional Cohen–Macaulay complex over $\mathbb{F}$. Suppose $\Delta$ satisfies the Link Condition with respect to an edge $e \in \Delta$. If $\mathcal{C}_\Delta(e)$ is Cohen–Macaulay over $\mathbb{F}$ of dimension $n-1$ and both $\mathrm{lk}_\Delta(e)$ and $\mathcal{C}_\Delta(e)$ are strong (respectively, almost strong) Lefschetz over $\mathbb{F}$, then so is $\Delta$.*

We note that since, e.g., by Reisner's criterion (see Theorem 1.28), the Cohen–Macaulay property is inherited by links, it is guaranteed that $\mathrm{lk}_\Delta(e)$ is Cohen–Macaulay. On the contrary, the contraction of an edge does not even need to be pure. Proposition 4.16 was proved in [Mur10, Proposition 3.2] for the strong Lefschetz property if $\mathbb{F}$ is an arbitrary infinite field of any characteristic (see also [BN10, Theorem 2.2] for the same result in characteristic zero). Since it is not entirely obvious, although reasonable to believe, that the proofs go through for the almost strong Lefschetz property, we sketch the main steps of the proof.

*Proof of Proposition 4.16.* Let $V$ be the vertex set of $\Delta$ and $e = \{a, b\} \in \Delta$, where $a < b$. Following [Mur10], we consider the shift operator

$$C_e(F) \;=\; \begin{cases} (F \smallsetminus \{b\}) \cup \{a\}, & \text{if } b \in F, \, a \notin F \text{ and } (F \smallsetminus \{b\}) \cup \{a\} \notin \Delta, \\ F, & \text{otherwise,} \end{cases}$$

which goes back to [EKR61], and set $\mathrm{shift}_e(\Delta) = \{C_e(F) : F \in \Delta\}$. Since the Link Condition holds for $e$, [Mur10, Lemma 2.1] implies that

$$\mathrm{shift}_e(\Delta) \;=\; \mathcal{C}_\Delta(e) \cup \{\{b\} \cup F \;:\; F \in a * \mathrm{lk}_\Delta(e)\}.$$

This implies that $\mathrm{shift}_e(\Delta) = \mathcal{C}_\Delta(e) \cup \mathrm{star}_\Delta(e)$ and, as a result, there is the exact sequence of $\mathbb{F}[x_v : v \in V]$-modules

$$0 \to \mathbb{F}[\mathrm{star}_\Delta(e)] \to \mathbb{F}[\mathrm{shift}_e(\Delta)] \to \mathbb{F}[\mathcal{C}_\Delta(e)] \to 0, \tag{4.1}$$

where the first map is given by multiplication with $x_b$. Since $\mathrm{lk}_\Delta(e)$ is $(n-3)$-Lefschetz, so is $\mathrm{star}_\Delta(e)$ (see, e.g., [KN09, Lemma 2.1]). Hence, there exist $\Theta = (\theta_1, \theta_2, \ldots, \theta_n)$ and a linear form $\omega \in \mathbb{F}[x_v : v \in V]$ such that $\Theta$ is an l.s.o.p. for $\mathbb{F}[\mathrm{star}_\Delta(e)]$, $\mathbb{F}[\mathrm{shift}_e(\Delta)]$ and $\mathbb{F}[\mathcal{C}_\Delta(e)]$ simultaneously and $\omega$ is an $(n-1)$- and $(n-3)$-Lefschetz element for $\mathcal{C}_\Delta(e)$ and $\mathrm{star}_\Delta(e)$, respectively, with respect to $\Theta$. Hence, from (4.1) we get the commutative diagram

$$
\begin{array}{ccccccc}
0 & \to & \mathbb{F}(\mathrm{star}_\Delta(e))_{\ell-1} & \to & \mathbb{F}(\mathrm{shift}_e(\Delta))_\ell & \to & \mathbb{F}(\mathcal{C}_\Delta(e))_\ell & \to & 0 \\
 & & \downarrow \omega^{n-3-2(\ell-1)} & & \downarrow \omega^{n-2\ell-1} & & \downarrow \omega^{n-2\ell-1} & & \\
 & & \mathbb{F}(\mathrm{star}_\Delta(e))_{n-2-\ell} & \to & \mathbb{F}(\mathrm{shift}_e(\Delta))_{n-1-\ell} & \to & \mathbb{F}(\mathcal{C}_\Delta(e))_{n-1-\ell} & \to & 0
\end{array}
$$

for $0 \le \ell \le \lfloor (n-1)/2 \rfloor$, where we have set $\mathbb{F}[\mathrm{star}_\Delta(e)]_{-1} = 0$ and $\mathbb{F}(\mathrm{star}_\Delta(e))$ for $\mathbb{F}[\mathrm{star}_\Delta(e)]/\Theta\mathbb{F}[\mathrm{star}_\Delta(e)]$, $\mathbb{F}(\mathrm{shift}_e(\Delta))$ for $\mathbb{F}[\mathrm{shift}_e(\Delta)]/\Theta\mathbb{F}[\mathrm{shift}_e(\Delta)]$, and $\mathbb{F}(\mathcal{C}_\Delta(e))$ for $\mathbb{F}[\mathcal{C}_\Delta(e)]/\Theta\mathbb{F}[\mathcal{C}_\Delta(e)]$.

Since the left and right vertical maps are injective by assumption, so is the middle map by the snake lemma. Thus, $\mathrm{shift}_e(\Delta)$ has the almost strong Lefschetz property. Moreover, since $\Delta$ satisfies the Link Condition with respect to $e$, we conclude from [Mur10, Lemma 2.2] that $I_{\mathrm{shift}_e(\Delta)}$ is an initial ideal of $I_\Delta$ with respect to a certain term order. Finally, $\Delta$ has the $(n-1)$-Lefschetz property by [Wie04, Proposition

2.9]. Wiebe's orginal result was for $\mathfrak{m}$-primary homogeneous ideals having the strong Lefschetz property. However, the same proof works in our setting. $\qquad\square$

**Definition 4.17.** Given a simplicial complex $\Delta$ and face $U = \{u_1, u_2, \ldots, u_n\} \in \Delta$, we say that $\Delta$ satisfies the *strong Link Condition* with respect to $U$ if

$$\mathrm{lk}_\Delta(F) \cap \mathrm{lk}_\Delta(G) \;=\; \mathrm{lk}_\Delta(F \cup G) \tag{4.2}$$

for all $F, G \subseteq U$ with $F \cap G = \varnothing$.

Note that, in this case, (4.2) holds for all (not necessarily disjoint) subsets $F, G \subseteq U$. The following technical lemma relates the strong Link Condition to the usual Link Condition.

**Lemma 4.18.** *Let $\Delta$ be a simplicial complex which satisfies the strong Link Condition with respect to the face $U = \{u_1, u_2, \ldots, u_n\}$. Then, $\mathcal{C}_\Delta(\{u_{n-1}, u_n\})$ satisfies the strong Link Condition with respect to $U \setminus \{u_n\}$.*

*In particular, all edges of $2^U$ can be contracted successively so that at each step, the Link Condition is satisfied with respect to the contracted edge.*

*Proof.* To simplify notation, we set $\Delta' = \mathcal{C}_\Delta(\{u_{n-1}, u_n\})$ and let $U' = U \setminus \{u_n\}$ be the vertex set of $\Delta'$. We consider disjoint sets $F, G \subseteq U'$ and observe that, by definition of $\Delta'$,

$$\mathrm{lk}_{\Delta'}(F) \;=\; \{H \in \mathrm{lk}_\Delta(F) \,:\, u_n \notin H\} \tag{4.3}$$
$$\cup \{(H \setminus \{u_n\}) \cup \{u_{n-1}\} \,:\, u_n \in H \in \mathrm{lk}_\Delta(F), \; u_{n-1} \notin H\}$$

if $u_{n-1} \notin F$, and

$$\mathrm{lk}_{\Delta'}(F) \;=\; \{H \in \mathrm{lk}_\Delta(F) \,:\, u_n \notin H\} \cup \mathrm{lk}_\Delta(F \cup \{u_n\}) \tag{4.4}$$

if $u_{n-1} \in F$. The inclusion

$$\mathrm{lk}_{\Delta'}(F \cup G) \;\subseteq\; \mathrm{lk}_{\Delta'}(F) \cap \mathrm{lk}_{\Delta'}(G)$$

holds trivially. To prove the reverse inclusion, we consider a face $H \in \mathrm{lk}_{\Delta'}(F) \cap \mathrm{lk}_{\Delta'}(G)$ and distinguish two cases.

Case 1: $u_{n-1} \notin F \cup G$. By Equation (4.3) for $\mathrm{lk}_{\Delta'}(F)$ and $\mathrm{lk}_{\Delta'}(G)$, four cases can occur. First, assume that $H \in \mathrm{lk}_\Delta(F) \cap \mathrm{lk}_\Delta(G)$. Then, by the strong Link Condition, $H \in \mathrm{lk}_\Delta(F \cup G)$ and hence $H \in \mathrm{lk}_{\Delta'}(F \cup G)$ by (4.3). Next, suppose that $H = (H' \setminus \{u_n\}) \cup \{u_{n-1}\}$ for some $H' \in \mathrm{lk}_\Delta(F) \cap \mathrm{lk}_\Delta(G)$ with $u_n \in H'$, $u_{n-1} \notin H$. Then, the strong Link Condition implies that $H' \in \mathrm{lk}_\Delta(F \cup G)$ and hence $H \in \mathrm{lk}_{\Delta'}(F \cup G)$ by (4.3). Finally, assume that $H \in \mathrm{lk}_\Delta(F)$ and that $H = (H' \setminus \{u_n\}) \cup \{u_{n-1}\}$ for some $H' \in \mathrm{lk}_\Delta(G)$ with $u_n \in H'$ and $u_{n-1} \notin H'$. Then, $F \cup H \in \Delta$ and $G \cup (H \setminus \{u_{n-1}\}) \cup \{u_n\} \in \Delta$. From the strong Link Condition, we conclude that $H \setminus \{u_{n-1}\} \in \mathrm{lk}_\Delta(F \cup G \cup \{u_n\})$, i.e., $(H \setminus \{u_{n-1}\}) \cup \{u_n\} \in \mathrm{lk}_\Delta(F \cup G)$.

This, together with (4.3) applied to $\mathrm{lk}_{\Delta'}(F\cup G)$, implies again that $H \in \mathrm{lk}_{\Delta'}(F\cup G)$. The remaining case follows by symmetry from the previous one.

**Case 2**: $u_{n-1} \in F \cup G$. Since $F \cap G = \varnothing$, we may assume without loss of generality that $u_{n-1} \in F$ and $u_{n-1} \notin G$. Since $H \in \mathrm{lk}_{\Delta'}(F)$ and $u_{n-1} \in F$, we must have $u_{n-1} \notin H$. From Equation (4.3), which applies to $\mathrm{lk}_{\Delta'}(G)$, and the fact that $u_{n-1} \notin H$ we conclude that $H \in \mathrm{lk}_{\Delta}(G)$. Two subcases can occur. Suppose first that $H \in \mathrm{lk}_{\Delta}(F)$. Then, the strong Link Condition implies that $H \in \mathrm{lk}_{\Delta}(F \cup G)$ and thus $H \in \mathrm{lk}_{\Delta'}(F \cup G)$ by Equation (4.4), applied to $\mathrm{lk}_{\Delta'}(F \cup G)$. Otherwise, $H \notin \mathrm{lk}_{\Delta}(F)$ and we must have $H \in \mathrm{lk}_{\Delta}((F \smallsetminus \{u_{n-1}\}) \cup \{u_n\})$ by (4.4). This implies that $H \cup (F \smallsetminus \{u_{n-1}\}) \cup \{u_n\} \in \Delta$. Since $H \cup G \in \Delta$, from the strong Link Condition we infer that $H \cup (F \smallsetminus \{u_{n-1}\}) \cup \{u_n\} \cup G \in \Delta$. Since $u_{n-1} \in F$, we conclude that $H \cup F \cup G \in \Delta'$ and so, once again, $H \in \mathrm{lk}_{\Delta'}(F \cup G)$. This completes the proof of the first statement.

For the second statement we note that if $\Delta$ satisfies the strong Link Condition with respect to $U$, then it also satisfies the Link Condition with respect to any edge of $2^U$. Hence, the claim follows from successive applications of the first statement. $\qquad\square$

## 4.2 Lefschetz properties of antiprism triangulations

This section aims to prove Theorem 4.1, i.e., to show that the antiprism triangulation of any shellable simplicial complex has the almost strong Lefschetz property over $\mathbb{R}$. From this we will infer that the $h$-vector of the antiprism triangulation of any Cohen–Macaulay simplicial complex is unimodal and will locate its peak.

We first show that the antiprism triangulation of the simplex $\sigma_n$ has the almost strong Lefschetz property over $\mathbb{R}$. The next lemma will be crucial.

**Lemma 4.19.** *Consider an $(n-1)$-dimensional simplex $2^V$ and a triangulation $\Delta$ of its boundary complex $\partial(2^V)$. Then, the antiprism $\Gamma_A(\Delta)$ satisfies the strong Link Condition with respect to the set of its interior vertices.*

*In particular, $\mathrm{sd}_A(\sigma_n)$ satisfies the strong Link Condition with respect to the set of its interior vertices.*

*Proof.* Set $V = \{v_1, v_2, \ldots, v_n\}$ and let $U = \{u_1, u_2, \ldots, u_n\}$ be the set of interior vertices of $\Gamma_A(\Delta)$, linearly ordered so that $\{u_i, v_i\} \notin \Gamma_A(\Delta)$ for every $i \in [n]$.

Let $E = \{u_i : i \in I\} \subseteq U$ for some $I \subseteq [n]$ be nonempty and let $\bar{E} = \{u_j : j \in [n] \smallsetminus I\}$ and $\bar{F} = \{v_j : j \in [n] \smallsetminus I\}$ be the faces of the simplices $2^U$ and $2^V$, respectively, which are complementary to $E$. Then, by definition of $\Gamma_A(\Delta)$,

$$\mathrm{lk}_{\Gamma_A(\Delta)}(E) = \Delta_A(\Delta_{\bar{F}}),$$

where the new vertices added for the $\Delta_A$ construction (see Remark 3.5) are the elements of $\bar{E}$ and $\Delta_{\bar{F}}$ is the restriction of $\Delta$ to the (proper) face $\bar{F} \in 2^V$. This directly implies that $\Gamma_A(\Delta)$ satisfies the strong Link Condition with respect to $U$. $\qquad\square$

**Definition 4.20.** Given a Cohen–Macaulay simplicial complex $\Delta$ over a field $\mathbb{F}$, we say that the contraction of an edge $e \in \Delta$ is *admissible over* $\mathbb{F}$ if $\Delta$ satisfies the Link Condition with respect to $e$ and $\mathrm{lk}_\Delta(e)$ is strong Lefschetz over $\mathbb{F}$.

The following proposition is, essentially, a consequence of Lemmas 4.18 and 4.19.

**Proposition 4.21.** *There exists a sequence of admissible edge contractions over* $\mathbb{R}$ *which transforms* $\mathrm{sd}_A(\sigma_n)$ *into the cone over its boundary. In particular,* $\mathrm{sd}_A(\sigma_n)$ *is almost strong Lefschetz over* $\mathbb{R}$, *if* $\partial(\mathrm{sd}_A(\sigma_n))$ *is strong Lefschetz over* $\mathbb{R}$.

*Proof.* As before, we let $V = \{v_1, \ldots, v_n\}$ and $U = \{u_1, \ldots, u_n\}$ be the vertices of $\sigma_n$ and the interior vertices of $\mathrm{sd}_A(\sigma_n)$, respectively. By Lemma 4.19, $\mathrm{sd}_A(\sigma_n)$ satisfies the strong Link Condition with respect to $U$. Thus, using Lemma 4.18, we can successively contract edges from $U$, each satisfying the Link Condition, until we reach a single vertex $u$. The resulting complex is clearly the cone $u * \partial(\mathrm{sd}_A(\sigma_n))$. If we can verify that the intermediate complexes, appearing in this sequence of contractions, are Cohen–Macaulay over $\mathbb{R}$ and that the links of the contracted edges are strong Lefschetz, then Proposition 4.16 implies that $\mathrm{sd}_A(\sigma_n)$ is almost strong Lefschetz, if so is the cone $u * \partial(\mathrm{sd}_A(\sigma_n))$.

To prove the missing statements, we use the fact that $\mathrm{sd}_A(\sigma_n)$ can be constructed from $\sigma_n$ by crossing operations on its faces, starting at the facet $V$ and moving to faces of lower dimension (see Section 3.2). From this it follows that the intermediate complexes can be constructed by first contracting the corresponding edges in the antiprism $\Gamma_A(\partial(2^V))$ and then performing crossing operations on its boundary faces. The antiprism $\Gamma_A(\partial(2^V))$ is a regular triangulation of $2^V$ and so is any subcomplex obtained from it by the performed edge contractions. Since, in addition, any crossing operation can be realized by a sequence of stellar subdivisions (see the proof of [BM87, Theorem 8]), which are well known to preserve regularity, we conclude that any intermediate complex in the sequence of edge contractions from $\mathrm{sd}_A(\sigma_n)$ to the cone $u * \partial(\mathrm{sd}_A(\sigma_n))$ is a regular triangulation of $2^V$ and, in particular, Cohen–Macaulay over $\mathbb{R}$. Moreover, the regularity of the intermediate complexes implies that the link of any interior edge that is contracted is a polytopal sphere and hence strong Lefschetz over $\mathbb{R}$ [Sta80]. Using Proposition 4.16, we conclude that $\mathrm{sd}_A(\sigma_n)$ is almost strong Lefschetz over $\mathbb{R}$, if so is $u * \partial(\mathrm{sd}_A(\sigma_n))$. By [KN09, Lemma 2.1], this is the case if $\partial(\mathrm{sd}_A(\sigma_n))$ is strong Lefschetz over $\mathbb{R}$. $\square$

The next statement suffices to conclude that $\mathrm{sd}_A(\sigma_n)$ has the almost strong Lefschetz property over $\mathbb{R}$.

**Proposition 4.22.** *The simplicial complex* $\partial(\mathrm{sd}_A(\sigma_n))$ *is combinatorially isomorphic to the boundary complex of a simplicial polytope. In particular, it is strong Lefschetz over* $\mathbb{R}$.

*Proof.* We use again the fact that $\partial(\mathrm{sd}_A(\sigma_n))$ can be constructed from $\partial\sigma_n$ by a sequence of crossing operations (see Section 3.2). As already mentioned, it was shown in the proof of Theorem 8 in [BM87] that every crossing operation can be expressed as a sequence of stellar subdivisions. Since those preserve polytopality, the first statement follows. The second follows from the first and Theorem 4.6.   $\square$

The next result follows by combining Propositions 4.21 and 4.22.

**Theorem 4.23.** *The simplicial complex* $\mathrm{sd}_A(\sigma_n)$ *is almost strong Lefschetz over* $\mathbb{R}$.

**Remark 4.24.** Since the restriction of the antiprism triangulation $\mathrm{sd}_A(\sigma_n)$ to a face $F$ of $\sigma_n$ is the antiprism triangulation of $2^F$, we can apply the edge contractions from the proof of Proposition 4.21 to the subdivided faces of $\partial(\mathrm{sd}_A(\sigma_n))$, ordered by decreasing dimension. Clearly, the simplicial complex obtained in this way is combinatorially isomorphic to the barycentric subdivision of $\partial\sigma_n$. Using similar arguments as in the proof of Proposition 4.21, one can show that all edge contractions are admissible. Indeed, let $\Delta'$ be the simplicial complex obtained from $\partial(\mathrm{sd}_A(\sigma_n))$ after $i$ edge contractions. Consider a face $F \in \sigma_n$ and let $\Delta'_F$ be the restriction of $\Delta'$ to $F$ (which is the subcomplex of $\Delta'$ consisting of all faces with carrier contained in $F$). If $G \in \Delta'_F$ is a face having all vertices in the interior of $\Delta'_F$, then it can easily be verified that

$$\mathrm{lk}_{\Delta'}(G) \;=\; \mathrm{lk}_{\Delta'_F}(G) * \{\{u_{H_1}, \ldots, u_{H_r}\} \;:\; F \subsetneq H_1 \subsetneq \cdots \subsetneq H_r \subsetneq [n]\}, \qquad (4.5)$$

where $u_H$ denotes the last interior vertex in the sequence of contractions of a face $H \in \sigma_n$. If $G$ is the edge to be contracted in $\Delta'$, then the previous equation, combined with the proof of Proposition 4.21, implies that $\Delta'$ satisfies the Link Condition with respect to $G$. We further note that the second complex on the right-hand side of (4.5) is isomorphic to the barycentric subdivision of $\mathrm{lk}_{\partial\sigma_n}(F)$, which itself is the barycentric subdivision of the boundary complex of an $(n-2-|F|)$-dimensional simplex. Thus, by the proofs of Propositions 4.13 and 4.21, both simplicial complexes on the right-hand side of (4.5) are strong Lefschetz over $\mathbb{R}$. From this fact and [BN10, Theorem 1.2 (i)], it follows that $\mathrm{lk}_{\Delta'}(G)$ is strong Lefschetz over $\mathbb{R}$.

The previous discussion shows that there exists a sequence of admissible edge contractions transforming $\partial(\mathrm{sd}_A(\sigma_n))$ into $\partial(\mathrm{sd}(\sigma_n))$. This provides another proof of the second statement of Proposition 4.22.

Another combinatorial property, that can be related to the Lefschetz property, is the so-called strongly edge decomposability.

**Definition 4.25.** (see [Mur10, Definition 1.1]) The boundary complex of a simplex and $\emptyset$ are *strongly edge decomposable*, and recursively, a simplicial complex $\Delta$ is *strongly edge decomposable*, if there exists an edge $\{i, j\} \in \Delta$ such that $\Delta$ satisfies the Link condition with respect to $\{i, j\}$ and both $\mathcal{C}_\Delta(ij)$ and $\mathrm{lk}_\Delta(ij)$ are strongly edge decomposable.

The existence of a sequence of admissible edge contractions mentioned before, together with the fact that we can use an analogous sequence of admissible edge contractions on the edge links, and that the barycentric subdivision of $\partial\sigma_n$ is strongly

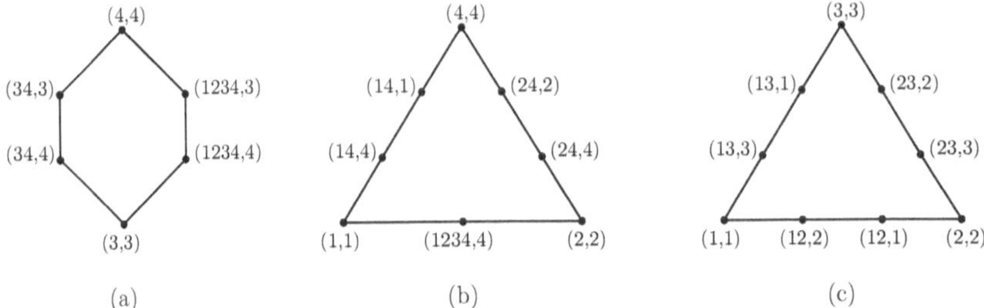

Figure 4.1: Links of the contracted edges $\{([4], i), ([4], i+1)\}$

edge decomposable, yield that $\partial\mathrm{sd}_A(\sigma_n)$ is strongly edge decomposable, too. Being strongly edge decomposable is a nice property for a simplicial complex in this context, since one can prove using Proposition 4.16, that strongly edge decomposable spheres also have the strong Lefschetz property.

**Proposition 4.26.** [Mur10, Corollary 3.5] *Strongly edge decomposable complexes are Cohen–Macaulay and have the strong Lefschetz property.*

**Example 4.27.** We sketch the procedure and the occuring edge-links described in Remark 4.24 at the example of the boundary of the antiprism triangulation of the 4-simplex $\partial(\mathrm{sd}_A(\sigma_5))$. We start by contracting all the interior edges in the subdivided facets, e.g., the edges $\{([4], i), ([4], i+1)\}$ for $1 \leq i \leq 3$ in the subdivided facet $[4] = \{1, 2, 3, 4\}$. The links of these edges in the contracted simplicial complex can be seen with labeled vertices in Figure 4.1.

The simplicial complex in (a) is the link of the edge $\{([4], 1), ([4], 2)\}$ in $\mathrm{sd}_A(\sigma_4)$, (b) is the link of $\{([4], 2), ([4], 3)\}$ after the first edge has been contracted, and (c) the link of $\{([4], 3), ([4], 4)\}$ after contracting the second edge. One can see, that these complexes again contain antiprism triangulations as subcomplexes, on which one can perform analogous sequences of edge contractions, as mentioned before.

Now we can contract the edges having all vertices in the interior of the subdivided lower-dimensional faces of $\sigma_4$, e.g., we consider the edges $\{([3], 1), ([3], 2)\}$ and $\{([3], 2), ([3], 3)\}$ in the subdivided face $\mathrm{sd}_A(2^{\{1,2,3\}})$ and we describe their links. For both edges, by equation (4.5), we have to take the join of the link of the edge in this subdivided face and $\{u_{\{1,2,3,4\}}, u_{\{1,2,3,5\}}\}$. In both cases, the link of the edge in the subdivided face is equal to the boundary of the 1-simplex. This yields the complexes in Figure 4.2, where complex (a) is the link of edge $\{([3], 1), ([3], 2)\}$ and (b) the link of edge $\{([3], 2), ([3], 3)\}$

With Proposition 4.22 at hand, the key observation to complete the proof of the first statement of Theorem 4.1 is that the proof of [KN09, Theorem 1.1], showing that the barycentric subdivision of any shellable simplicial complex is almost strong

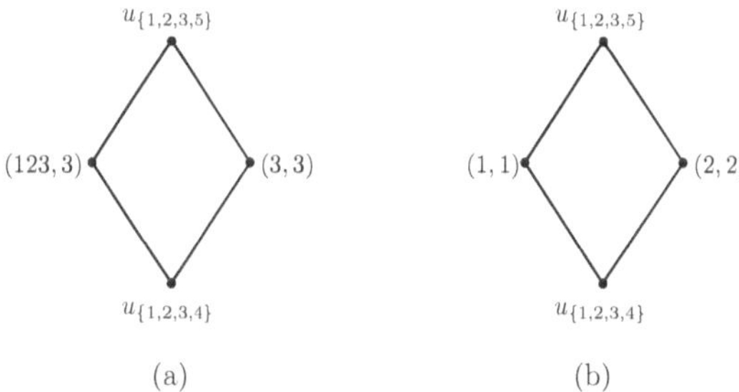



Figure 4.2: Links of the contracted edges $\{([3],i),([3],i+1)\}$

Lefschetz over an infinite field (in particular, over $\mathbb{R}$), works for every uniform triangulation which fulfills this property for simplices. For the sake of completeness, we provide a sketch of the proof.

**Theorem 4.28.** *The complex* $\mathrm{sd}_A(\Delta)$ *is almost strong Lefschetz over* $\mathbb{R}$ *for every shellable simplicial complex* $\Delta$.

*Proof.* Let $\dim(\Delta) = n - 1$, as usual. The proof proceeds by double induction on $n$ and the number of facets of $\Delta$. At the base of the induction, either $\Delta$ consists only of vertices, in which case there is nothing to show, or $\Delta$ is a simplex, in which case the result follows from Proposition 4.22.

For the induction step we assume that $n \geq 2$, let $V$ be the vertex set of $\Delta$ and let $A = \mathbb{R}[x_v : v \in V]$. Consider a shelling $G_1, G_2, \ldots, G_m = G$ of $\Delta$ and set $\widetilde{\Delta} := \langle G_1, \ldots, G_{m-1} \rangle$ and $\tau := \widetilde{\Delta} \cap 2^G$. There is the following exact sequence of $A$-modules:

$$0 \to \mathbb{R}[\mathrm{sd}_A(\Delta)] \to \mathbb{R}[\mathrm{sd}_A(\widetilde{\Delta})] \oplus \mathbb{R}[\mathrm{sd}_A(2^G)] \to \mathbb{R}[\mathrm{sd}_A(\tau)] \to 0. \qquad (4.6)$$

One now chooses generic linear forms $\Theta = \theta_1, \theta_2, \ldots, \theta_n$ so that $\Theta$ is an l.s.o.p. for $\mathbb{R}[\mathrm{sd}_A(\Delta)]$, $\mathbb{R}[\mathrm{sd}_A(\widetilde{\Delta})]$ and $\mathbb{R}[\mathrm{sd}_A(2^G)]$ simultaneously and $\theta_1, \theta_2, \ldots, \theta_{n-1}$ is an l.s.o.p. for $\mathbb{R}[\mathrm{sd}_A(\tau)]$. Dividing out by $\Theta$ in (4.6) gives rise to an exact sequence

$$\mathrm{Tor}_1(\mathbb{R}[\mathrm{sd}_A(\tau)], A/\Theta) \xrightarrow{\delta} \mathbb{R}(\mathrm{sd}_A(\Delta)) \to \mathbb{R}(\mathrm{sd}_A(\widetilde{\Delta})) \oplus \mathbb{R}(\mathrm{sd}_A(2^G)) \to \mathbb{R}(\mathrm{sd}_A(\tau)) \to 0,$$

where we have written $\mathbb{R}(\mathrm{sd}_A(\Delta))$ for $\mathbb{R}[\mathrm{sd}_A(\Delta)]/\Theta$ and similarly for $\mathbb{R}(\mathrm{sd}_A(\widetilde{\Delta}))$, $\mathbb{R}(\mathrm{sd}_A(\tau))$ and $\mathbb{R}(\mathrm{sd}_A(2^G))$. Next, one shows that $\mathrm{Tor}_1(\mathbb{R}[\mathrm{sd}_A(\tau)], A/\Theta)_i = 0$ for $0 \leq i \leq \lfloor \frac{n-2}{2} \rfloor$. This is done exactly as in the proof of [KN09, Theorem 1.1]. Since all maps in the previous exact sequence preserve the grading, one gets the commutative diagram

$$\mathbb{R}(\mathrm{sd}_A(\Delta))_i \quad \to \quad \mathbb{R}(\mathrm{sd}_A(\widetilde{\Delta}))_i \oplus \mathbb{R}(\mathrm{sd}_A(2^G))_i$$

$$\downarrow \omega^{n-2i-1} \qquad\qquad \downarrow (\omega^{n-2i-1}, \omega^{n-2i-1})$$

$$\mathbb{R}(\mathrm{sd}_A(\Delta))_{n-1-i} \quad \to \quad \mathbb{R}(\mathrm{sd}_A(\widetilde{\Delta}))_{n-1-i} \oplus \mathbb{R}(\mathrm{sd}_A(2^G))_{n-1-i}$$

where $\omega$ is a degree one element in $A$. The induction hypothesis implies that the multiplication map on the right-hand side is injective for a generic $\omega$. One concludes that the multiplication

$$\omega^{n-2i-1} : \ \mathbb{R}(\mathrm{sd}_A(\Delta))_i \to \mathbb{R}(\mathrm{sd}_A(\Delta))_{n-1-i}$$

is injective for $0 \leq i \leq \lfloor \frac{n-2}{2} \rfloor$ and the proof follows. $\qquad\square$

To complete the proof of Theorem 4.1, we need the following properties of the numbers $p_A(n, k, j)$, discussed in Section 3.5.

**Lemma 4.29.** *Let $n, k, j \in \mathbb{N}$ with $k, j \leq n$.*

*(a) $p_A(n, k, j) = p_A(n, n - k, n - j)$,*

*(b)*

$$p_A(n, k, 0) \leq p_A(n, k, 1) \leq \cdots \leq p_A(n, k, \lfloor n/2 \rfloor)$$

*and*

$$p_A(n, k, n) \leq p_A(n, k, n - 1) \leq \cdots \leq p_A(n, k, \lceil n/2 \rceil).$$

*Proof.* Although part (a) follows from [Ath20b, Proposition 4.6 (a)], we provide a direct combinatorial proof as follows. Let $Q(n, k, j)$ be defined as in the proof of Proposition 3.27, so that $p_A(n, k, j) = |Q(n, k, j)|$. Also, let $\widetilde{Q}(n, n - k, n - j)$ be the set of triples defining $Q(n, n - k, n - j)$, except that the set $S$ which is partitioned satisfies $\{k + 1, \ldots, n\} \subseteq S \subseteq [n]$, instead of $[n - k] \subseteq S \subseteq [n]$, and that the monochromatic block, if present, must be contained in $\{k + 1, \ldots, n\}$. Since, clearly, $|\widetilde{Q}(n, n - k, n - j)| = p_A(n, n - k, n - j)$, to prove (a) it suffices to find a bijection from $Q(n, k, j)$ to $\widetilde{Q}(n, n - k, n - j)$. Given a triple in $Q(n, k, j)$, consisting of an ordered partition of $[k] \subseteq S \subseteq [n]$ and suitable coloring of the elements of $S$, we construct a triple in $\widetilde{Q}(n, n - k, n - j)$ as follows. We first switch the colors of all elements of $S$ from white to black and vice versa. If the first block was monochromatic, we delete it from the partition. The block $[n] \smallsetminus S$, if nonempty, is then added to the constructed ordered partition as its new first block, with all its elements colored black. We leave to the reader to verify that this process gives a well defined map. The inverse map can be constructed by the same procedure, applied to the triples in $\widetilde{Q}(n, n - k, n - j)$.

For part (b), we note that the proof of [KN09, Corollary 4.4] works, with the symmetry of [BW08, Lemma 2.5] replaced by that of part (a). $\qquad\square$

Theorem 4.28 has the following numerical consequences for the $h$-vector of the antiprism triangulation of a Cohen–Macaulay complex.

**Corollary 4.30.** *Let $\Delta$ be a Cohen–Macaulay simplicial complex of dimension $n-1$ and let $g(\mathrm{sd}_A(\Delta)) = (1, h_1(\mathrm{sd}_A(\Delta)) - h_0(\mathrm{sd}_A(\Delta)), \ldots, h_{\lfloor n/2 \rfloor}(\mathrm{sd}_A(\Delta)) - h_{\lfloor n/2 \rfloor - 1}(\mathrm{sd}_A(\Delta)))$.*

*(a) $g(\mathrm{sd}_A(\Delta))$ is an $M$-sequence.*

*(b) $h(\mathrm{sd}_A(\Delta))$ is unimodal. The peak is at position $n/2$, if $n$ is even, and at position $(n-1)/2$ or $(n+1)/2$, if $n$ is odd.*

*(c) $h_i(\mathrm{sd}_A(\Delta)) \leq h_{n-1-i}(\mathrm{sd}_A(\Delta))$ for all $0 \leq i \leq \lfloor (n-2)/2 \rfloor$.*

*Proof.* Parts $(a)$ and $(c)$ follow by standard arguments used when one works with Lefschetz properties, similar as described in Section 4.1.2. Since the $h$-vector of the antiprismatic triangulation of a simplicial complex $\Delta$ is a function of $h(\Delta)$, with Theorem 1.33 we can assume that $\Delta$ is shellable. $\Delta$ is almost strong Lefschetz by Theorem 4.1, hence for a generic l.s.o.p. $\Theta$ and a generic degree one element $\omega$ the multiplication

$$\omega^{n-1-2i} : (\mathbb{F}[\mathrm{sd}_A(\Delta)]/\Theta\mathbb{F}[\mathrm{sd}_A(\Delta)])_i \to (\mathbb{F}[\mathrm{sd}_A(\Delta)]/\Theta\mathbb{F}[\mathrm{sd}_A(\Delta)])_{n-1-i} \qquad (4.7)$$

is an injection for every $0 \leq i \leq \lfloor \frac{n-2}{2} \rfloor$. It follows, that also the multiplication

$$\omega : (\mathbb{F}[\Delta]/\Theta\mathbb{F}[\Delta])_i \to (\mathbb{F}[\Delta]/\Theta\mathbb{F}[\Delta])_{i+1}$$

is an injection. Since $\Delta$ is Cohen–Macaulay, by Theorem 4.11 we have $h_i(\mathrm{sd}_A(\Delta)) = \dim(\mathbb{F}[\mathrm{sd}_A(\Delta)]/\Theta\mathbb{F}[\mathrm{sd}_A(\Delta)])_i$. This yields

$$g_i(\mathrm{sd}_A(\Delta)) = \dim(\mathbb{F}[\mathrm{sd}_A(\Delta)]/(\Theta, \omega)\mathbb{F}[\mathrm{sd}_A(\Delta)])_i$$

for $0 \leq i \leq \lfloor \frac{n}{2} \rfloor$, showing that $g(\mathrm{sd}_A(\Delta))$ is an $M$-sequence.

Part $(c)$ follows by the fact that the multiplication in (4.7) is injective, and together with $h_i(\mathrm{sd}_A(\Delta)) = \dim(\mathbb{F}[\mathrm{sd}_A(\Delta)]/\Theta\mathbb{F}[\mathrm{sd}_A(\Delta)])_i$ we get $h_i(\mathrm{sd}_A(\Delta)) \leq h_{n-1-i}(\mathrm{sd}_A(\Delta))$ for all $0 \leq i \leq \lfloor \frac{n-2}{2} \rfloor$

For part (b), the proof of [KN09, Corollary 4.7] works, with Lemma 4.29 replacing [KN09, Corollary 4.4 (ii)]. $\qquad\square$

We conclude this chapter by recording the following properties of the Stanley-Reisner ring of the antiprism triangulation of any simplicial complex.

**Proposition 4.31.** *Let $\Delta$ be a simplicial complex.*

*(a) $\dim(\mathbb{F}[\mathrm{sd}_A(\Delta)]) = \dim(\mathbb{F}[\Delta])$.*

*(b) $\mathrm{depth}(\mathbb{F}[\mathrm{sd}_A(\Delta)]) = \mathrm{depth}(\mathbb{F}[\Delta])$.*

*(c)*

$$\mathrm{reg}(\mathbb{F}[\mathrm{sd}_A(\Delta)]) = \begin{cases} \dim(\Delta), & \text{if } \widetilde{H}_{\dim(\Delta)}(\Delta; \mathbb{F}) = 0, \\ \dim(\Delta) + 1, & \text{if } \widetilde{H}_{\dim(\Delta)}(\Delta; \mathbb{F}) \neq 0. \end{cases}$$

*Proof.* Part (a) is clear, since $\dim \mathrm{sd}_{\mathcal{A}}(\Delta) = \dim(\Delta)$. Part (b) follows from [Mun84, Theorem 3.1], since $\Delta$ and $\mathrm{sd}_{\mathcal{A}}(\Delta)$ have homeomorphic geometric realizations. Part (c) follows from an application of Hochster's formula [MS05, Corollary 5.12]; one can also mimick the detailed argument for barycentric subdivision given in the proof of [KW08, Proposition 2.6]. $\qquad\square$

# 5 Further directions

This chapter concludes with some comments, open problems and further directions for research.

1. The question asking which uniform triangulations transform $h$-polynomials with nonnegative coefficients into polynomials with only real roots was raised in [Ath20b]. This property has been verified for several examples, including the prototypical one of the barycentric subdivision [BW08], and is conjectured in this paper for the antiprism triangulation. We believe that this property is not uncommon among uniform triangulations.

For instance, consider any word $w = w_1 w_2 \cdots w_d$ with $w_i \in \{a, b\}$ for every $i \in [d]$ and let $\Delta$ be a $d$-dimensional simplicial complex. Let $\mathrm{sd}_w(\Delta)$ be the triangulation of $\Delta$ defined inductively as follows. Assume that all faces of $\Delta$ of dimension less than $j$ have been triangulated, for some $j \in [d]$. Then, triangulate each $j$-dimensional face $F \in \Delta$ by the antiprism construction, if $w_j = a$, and by the coning construction, if $w_j = b$, over the already triangulated boundary of $F$. By applying this process for $j = 1, 2, \ldots, d$, in this order, we get a triangulation $\mathrm{sd}_w(\Delta)$ of $\Delta$ which coincides with $\mathrm{sd}_{\mathcal{A}}(\Delta)$, when $w = aa \cdots a$, and with $\mathrm{sd}(\Delta)$, when $w = bb \cdots b$. It seems plausible that this triangulation has the same property for every $w$, but it is not easy to deduce such a statement from the results of [Ath20b] and this paper. Example 3.9 corresponds to the word $w = bb \cdots ba$.

Another way generalizing the antiprism triangulation could be to think of *partial antiprism triangulations*, similar to the approach in [AW18], where the partial barycentric subdivision was studied.

**Definition 5.1.** For $1 \leq l \leq n$, the *$l$-th partial antiprism triangulation* $\mathrm{sd}_{\mathcal{A}}^l(\sigma_n)$ of the simplex $\sigma_n$ is defined to be the complex obtained by performing the crossing operation starting from facets moving to faces of lower dimensions down to dimension $n - l$. For $l = 0$, we set $\mathrm{sd}_{\mathcal{A}}^0(\sigma_n) = \sigma_n$.

We see, that $l = n$ yields the antiprism triangulation defined in the previous section, i.e., $\mathrm{sd}_{\mathcal{A}}^n(\sigma_n) = \mathrm{sd}_{\mathcal{A}}(\sigma_n)$, and $\mathrm{sd}_{\mathcal{A}}^1(\sigma_3)$ is the complex shown in Figure 3.1.

Adapting the description of the faces of the antiprism triangulation in Section 2.2, we can describe the faces of the $l$-th partial antiprism triangulation in the following way: The vertices of $\mathrm{sd}_{\mathcal{A}}^l(\sigma_n)$ are the pointed faces $(F, v)$ of $\sigma_n$ with $\dim(F) \geq n - l$ or $\dim(F) = 0$. A set $\{(F_0, v_0), \ldots, (F_k, v_k)\}$ is a $k$-face if and only if all the $v_i$ are distinct, and the following properties hold for some $-1 \leq r \leq n - l,$:

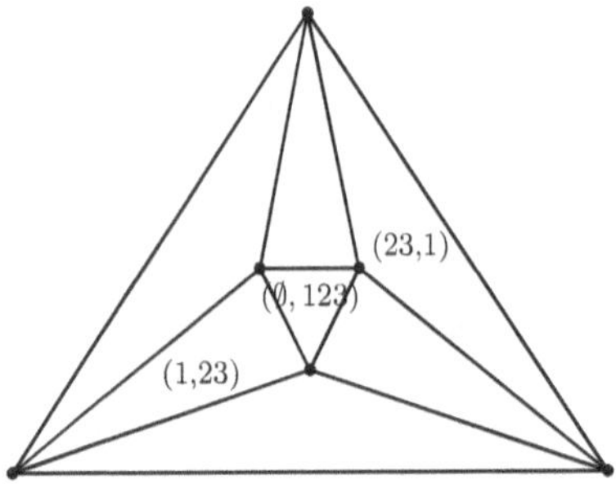


Figure 5.1: $\mathrm{sd}^1_{\mathcal{A}}(\sigma_3)$, with some labeled facets

1. $F_{r+1} \subseteq \cdots \subseteq F_k$ is a flag in $\sigma_n$, and $\dim(F_i) = 0$ and $F_i \subseteq F_{r+1}$ for $0 \leq i \leq r$,

2. if $F_i \subsetneq F_j$, then $v_j \notin F_i$ for $i, j \geq r + 1$.

Alternatively we can also describe the faces by a variant of multi-pointed set partitions, that we will also just call multi-pointed set partitions in the following. For a set $S$, a multi-pointed set partition of $S$ is a pair of set systems $(B_0, B_1, \ldots, B_r)$ and $(C_0, C_1, \ldots, C_r)$, such that $\bigcup_{i=0}^{r} B_i = S$ and $\bigcup_{i=0}^{r} C_i \subseteq S$, $C_i \subseteq B_i$ and $C_i \neq \emptyset$ for $1 \leq i \leq r$ and $C_0 = B_0$. We call $|\bigcup C_i|$ the weight of the multi-pointed set partition. In particular, the only difference to the earlier defined multi-pointed set partitions is the possibility of having an empty first block in the partition.

Now the $(k-1)$-dimensional faces of $\mathrm{sd}^l_{\mathcal{A}}(\sigma_n)$ correspond to the multi-pointed set partitions of weight $k$ with

1. $|B_0| = |C_0| \leq n - l$,

2. if $r \geq 1$, then $|B_0 \cup B_1| \geq n - l + 1$

As for the standard antiprism triangulation, the facets of the $l$-th partial antiprism triangulation can be described combinatorially using set partitions with the following properties: Let $(B_0, B_1, \ldots, B_r)$ be an ordered set partition of $[n]$, such that

1. $|B_0| \leq n - l$,

2. $r \leq l$.

3. if $r \geq 1$, then $|B_0 \cup B_1| \geq n - l + 1$

Figure 5.1 shows the example of the first partial antiprism triangulation of the simplex $\sigma_3$ with some labeled facets. For these subdivisions we can ask similar questions as for the standard antiprism triangulation about the $h$-vector transformation and the real-rootedness of the $h$-polynomial.

2. The symmetric polynomials $\ell_{\mathcal{A}}(\sigma_n, x)$ and $\bar{p}_{\mathcal{A}}(\sigma_n, x)$ were conjectured to be $\gamma$-positive in Section 3.3. It is an interesting (and possibly challenging) open problem to find explicit combinatorial interpretations of the corresponding $\gamma$-coefficients. Similar remarks apply to the symmetric polynomials $h_{\mathcal{A}}(\partial \sigma_n, x)$ and $\theta_{\mathcal{A}}(\sigma_n, x) = h_{\mathcal{A}}(\sigma_n, x) - h_{\mathcal{A}}(\partial \sigma_n, x)$, already shown to be real-rooted, and hence $\gamma$-positive, in Section 3.3.

3. The local $h$-polynomial of the barycentric subdivision of any CW-regular subdivision of the simplex was shown to be $\gamma$-positive in [JKMS19]. Does this hold if 'barycentric subdivision' and 'CW-regular subdivision' are replaced by 'antiprism triangulation' and 'CW-regular simplicial subdivision', respectively?

4. The $h$-polynomial of the barycentric subdivision of any doubly Cohen–Macaulay level simplicial complex and the barycentric subdivision of any triangulation of a ball were shown to have a nonnegative real-rooted symmetric decomposition in [BS19, Section 5] and [Ath20b, Section 8], respectively. Do these statements hold if 'barycentric subdivision' is replaced by 'antiprism triangulation'?

5. It is natural to ask whether, in addition to the numerical consequences of Theorem 4.1, the algebraic part of the statement holds for the antiprism triangulation of any Cohen–Macaulay complex as well.